Organization-Wide Physical Asset Management

A Systems Approach

Organization-Wide Physical Asset Management

A Systems Approach

Dharmen Dhaliah

INDUSTRIAL PRESS, INC.

Industrial Press, Inc.
32 Haviland Street, Suite 3
South Norwalk, Connecticut 06854
Phone: 203-956-5593
Toll-Free in USA: 888-528-7852
Email: info@industrialpress.com

Author: Dharmen Dhaliah
Title: Organization-Wide Physical Asset Management: A Systems Approach
Library of Congress Control Number: 2019951359

ISBN (print): 978-0-8311-3652-9
ISBN (ePDF): 978-0-8311-9552-6
ISBN (ePUB): 978-0-8311-9553-3
ISBN (eMOBI): 978-0-8311-9554-0

Editorial Director/Publisher: Judy Bass
Copy Editor: Elise Davies
Compositor: Patricia Wallenburg, TypeWriting
Proofreader: Judy Duguid
Indexer: Karin Arrigoni

books.industrialpress.com
ebooks.industrialpress.com
10 9 8 7 6 5 4 3 2

This book is dedicated to asset management practitioners,
senior leadership members, and most specifically
to aspiring asset management professionals
including graduate, undergraduate, and post graduate students
looking for advice to get into this amazing field
of physical asset management.

Good luck to everyone in their Organization-Wide
Physical Asset Management journey!

CONTENTS

PREFACE

The electric light did not come from
the continuous improvement of candles.
—Oren Harari

If you read my first book, *Physical Asset Management: An Organizational Challenge*, you may have been wondering what comes next. You may have several questions or concerns regarding physical asset management challenges in organizations and how to resolve them. You may have questions such as "What if we no longer did what we do now?"—not with the intention of abandoning current physical asset management activities, but as a way to take advantage of existing strengths and new opportunities to resolve the existing challenges. What are our alternatives in terms of approach to get different (better) results? If you have not read the book, I recommend that you grab a copy and come up with some questions yourself. This will help frame the challenges your organization is actually facing to implement a holistic physical asset management philosophy.

In this book we will continue to elaborate more on the subject of holistic physical asset management and the approach that organizations need to take. Hopefully some, if not all, of your questions will get answered, while new ones keep cropping up. There is no magic pill or straight answer to resolve the challenges. But understanding them and being more prepared to face them will definitely help.

We previously looked at the asset management field and how its "new" expectations fit in organizations along with the challenges of implementing a holistic approach.

Physical Asset Management: An Organizational Challenge was published in October 2016. One of the main reasons I wrote the book is that there is a pressing need for radical changes in organizations in order to overcome the challenges in implementing a holistic approach to managing physical assets.

After several years working in both the public and private sectors, in different industries, and, most importantly, in different functional areas, I have witnessed firsthand how physical assets are managed from different perspectives. Over my years of teaching and interaction with many asset management practitioners, I have heard and observed how organizations struggle to balance cost, performance, and risk with regard to the management of physical assets. The question is, what should we do about it, or—even better—what are we doing about it?

Just "bitching and moaning about it," as E. Goldratt likes to put it when he talks about the theory of constraints, will not help overcome the challenges.

It's time for us asset management practitioners to do something different, even though it will be a daunting undertaking. We need to take the bull by its horns and lead a radical transformation in organizations to be able to meet the requirements of asset management as specified in its definition as per the ISO 55000 standards. There are so many standards, documentations, conceptual models, articles, and other resources out there teaching us how to implement asset management. Almost every week we hear about webinars, conferences, or workshops being held to train and certify staff in asset management. But is that enough? When we talk about asset management in an organization, we are talking about a system—not a project or a program. And when we talk about a system, we need a systemic approach to be successful.

Dr. W. Edwards Deming explained that any organization is a system, that is, a network of interdependent components that share a common goal. In his theory of profound knowledge (TPK), Deming outlines four elements which are crucial to manage such a network of interdependencies: appreciation for a system, knowledge about variation, theory of knowledge, and knowledge of psychology. The four elements of Deming's theory tie into his popular and globally adopted fourteen points for total quality management. Deming believed his theory of profound knowledge was so important, because it would help the transformation process within organizations.

To realize maximum value from the transformation process of such a complex system as an organization, it is very important to understand the value chain(s) that exist. Value chain analysis is essential to demonstrate the causes and effects that activities have on each other to generate best value in the process. Michael Porter's value chain analysis is a great tool that can be applied in the development of an asset management system.

During the process of realizing value, any system will have constraints. The theory of constraints developed by Dr. Eliyahu M. Goldratt can be helpful in improving the asset management system by taking a methodical approach using its five focusing steps to eliminate those constraints and achieve the system's goals.

In this book, we look at asset management in an organization as a system of a complex network of interdependencies, consisting of interrelations and interactions. This innovative approach to asset management is inspired from when I was completing my MBA program at Heriot Watt University, Edinburgh Business School. One of the courses in the program was Strategic Risk Management, and the approach to the enterprise-wide risk management system (EWRMS) caught my attention. The goals, challenges, and principles of EWRMS are so similar to those of organization-wide physical asset management that they can be readily applied to building an effective asset management system in any organization. The book elaborates on this approach and how it applies to the complex network of interdependencies existing in an asset management system.

We also look at the system from an organizational perspective and apply systemic thinking to analyze the interdependencies. Throughout the book we are going to use Dr. Deming's theory of profound knowledge to build the asset management philosophy/approach, Michael Porter's value chain analysis to develop the asset management system to realize value, and Dr. Goldratt's theory of constraints to continuously improve the system by eliminating constraints. This will help the transformational process desperately required to implement an organization-wide physical asset management.

As Dr. Deming puts it, "Everyone doing his best is not the answer. It is first necessary that people know what to do. Drastic changes are required. The first step in the transformation is to learn how to change." We need to do something different by building on the tons of tools, methodologies, and standards available to transform organizations. It is time to reach out to all the asset management practitioners and everyone else in the organization who deals with physical assets in one way or the other—from maintenance professionals to the project manager who delivers capital projects, to the operator who operates the asset, to the decision maker who makes investment decisions, to the financial analyst who prepares the capital budget, to the accountant who puts together the physical asset valuation report, and to others. The message must be "Business as usual will not cut it anymore. It's time for a paradigm shift!"

It is time we bring everyone together to understand the holistic approach of managing physical assets over their entire lifecycle; to understand how our actions/inactions or decisions can impact the effectiveness of the process; to understand how working together will make better use of the resources at our disposal; and to understand how harmonization and collaboration can help make better decisions.

There are conferences and training programs taking place every day globally on the subject of asset management, mostly concerned with the strategies and tools. They tell you that there is a problem, but they don't tell you how to resolve it. This book attempts to provide some solutions to the problem while applying some of the strategies and tools.

Managing physical assets in a chaotic style is not good enough. In other words, it's time for all of us to embrace the organization-wide approach of managing physical assets and start the transformation process!

This should not be a moment but a movement—a movement to transform an organization. For that to happen we have to face the "beast"—the complex system of interdependencies in organizations.

ACKNOWLEDGMENTS

Writing a book is really a collaborative effort to effectively gather thoughts and perspectives from different sources. Sources of inspiration are important, and this work is a synergistic product of many minds, different philosophies, and various brilliant expertise.

First of all, I would like to bow to the great work and heritage that Dr. Deming, Dr. Goldratt, and Mr. Michael Porter have left us. Their contribution has set a foundation that we practitioners in every industry are still using to advance our respective fields. In the book *Quality Involvement Flow: The Systemic Organization* (Lepore, Montgomery, and Siepe), the authors have done a brilliant job in combining Deming's and Goldratt's works. Thank you for providing that perspective.

Next I would like to praise and acknowledge the inspiring work of authors Dr. William Wallace, Neil McClure, and Shai Davidov in the Strategic Risk Management Course from Heriot-Watt University Edinburgh Business School's online MBA programme. A big thank you to Heriot-Watt University for allowing me to use the concepts in the course text and apply them to physical asset management.

A huge mention to Dr. Arnold Yuan, Associate Professor at Ryerson University, who provided me with some invaluable insights on his teaching and research experience.

A big thank you to all my colleagues, peers, and students and to all my network with whom I have interacted at conferences, on social media, or in seminars/webinars. One way or another, they all inspired me in writing this book—their questions, perspectives, experience, challenges, successes, and struggles.

A special mention to the fresh graduates and professionals interested in getting in the asset management field. You are one of the main reasons I wrote this book. I hope you will enjoy it and wish you all the best in asset management.

A big thank you to Darshana Appadoo and Balraj Appadu for their creative skills in putting together all the graphics in the book. It was a long and tedious task.

Last but not least I am extremely grateful to my family—my wife, Mitrini, and my sons, Yuvneesh and Yaneesh. They have inspired and encouraged me all the way, with helpful feedback and advice as I was putting the chapters together. Thank you for keeping the faith and for your endless patience, encouragement, and love.

INTRODUCTION

On a misty cold morning, Darren Ordlan is shivering in his wetsuit as he stands among the other participants waiting for their wave to start. He is about to take on his first Ironman 70.3 race in Lake Placid, New York. This was a goal he set for himself when he turned fifty years old. It is not going to be easy, and he knows that. He is jumping on the spot, while shaking his arms and legs to fight the cold while simultaneously trying to relieve the stress that is making his stomach churn. But there is no going back, and he knows that too. He has trained relentlessly over the last few years for this specific moment and made lots of sacrifices along the way. These sacrifices did not only impact him, but his wife, Debbie, and his two sons, Rémi and Lucas, as well, who supported him all the way. They all made the trip to Lake Placid with him—a beautiful road trip of around six hours with fascinating views. Darren told Debbie not to wake up early to come with him for the start preparation, as it would be too cold in the morning. He did tell her and the boys to come cheer him on at the starting line. When that time came, Darren could not find them in the huge crowd of spectators along the beautiful shore of Lake Placid. The participants standing together on the beach keep moving slowly as they hear the sound of the siren. Although it is cold this morning, the water itself is quite warm when Darren gets in to do his warm-up. He could see the vapor fumes coming off the lake's surface creating a majestic scene usually seen in movies. Now it is time to stop his wandering mind and focus on the race and what lies ahead of him for the next seven to eight hours. Darren needs to execute all the strategies and plans that he has worked on for the last few years to the best of his abilities.

"Dad!" Darren hears a voice calling him from afar. He looks around and finally spots Debbie, Rémi, and Lucas waving at him from among the crowd two or three rows away. They made it, braving the cold morning, and Darren is pleased to see them there supporting him. It gives him more confidence and purpose for what he is about to undertake.

"Wwwuuummppp!" goes the loud siren, and everybody around him starts running. He understands right away—this is it! He starts running, following the crowd the best he can, as there is a lot of jostling happening. He splashes into the water and starts to swim. It is a rough start with so many people around, arms flailing and legs kicking at him. Darren knows he has to keep his cool and stay focused for the next few minutes to steady himself. After a few strokes, he finally gets into a good rhythm, and his breathing also starts to settle down. It is a long swim, and he needs to manage his energy well. Darren feels very good taking deep breaths and letting his body rotate and glide in the clear water. He is now enjoying his swim at a good pace. But it does not last long.

"Drrriiinnnng!" His alarm clock starts to ring, and he reaches out to switch it off.

What a nice dream he was having. For the last four years he regularly relives this Lake Placid Ironman race even though he has completed multiple similar races since then at other locations. The Lake Placid Ironman race remains a special race, first because it was his first race; second, because of its magical location; and third, because it was such a great accomplishment in the sense that he proved "nothing is impossible."

Darren peeks at the alarm clock, and he cannot believe it is already six o'clock. "Back to reality," he says to himself.

It is Saturday, and he needs to go for his swimming session, which starts at 7:00, and then he has to meet up with Jerry Detunam at 8:30 am. His schedule is tight, and, unsurprisingly, he is already late! In a matter of minutes, Darren gets ready and rushes downstairs. He puts his swimming stuff into his bag and heads to the garage.

While driving to the swimming pool, Darren thinks about Jerry and how he can help him. He has known Jerry for a few years now, since they met in one of the classes on physical asset management that he was teaching. Jerry is the new asset manager at Mouroum Inc., a global leader in the food manufacturing industry. Lately, Jerry has been calling Darren almost daily, as he has been struggling in his job. Jerry wanted to book an appointment with Darren as soon as possible to discuss asset management and the issues he is facing at work. Over the phone he sounded somewhat stressed out about his new job, and he wanted to get some advice.

Darren has been in the field of managing physical assets for over thirty years in different industries of both the private and public sectors. His career has taken him through the tactical aspects of managing physical assets to the strategic approach of physical asset management. He has been successful not only in developing physical asset management strategies and plans but also in actually implementing them successfully in organizations and reaping the benefits while witnessing firsthand the challenges that come with it.

Two years ago he decided to go for a change and take on a new challenge. He took early retirement from his high-paying job in asset management to fulfill his dream of doing research work and completing his PhD in the field of asset management. Debbie was against his decision to take early retirement, but Darren's love for asset management and teaching, as well as his desire to leave a legacy, was so strong that it was virtually impossible to make him change his mind. He joined Ryerson University as an adjunct professor. As an adjunct professor, he is not on the tenure track, but he teaches courses just like tenured professors do. He knows that the career path to earning his PhD and becoming a tenured professor is a rocky one. He also knows the job is not guaranteed, and compensation is quite low. That is why he also takes on consultancy work in asset management, when the opportunity arises, to help make ends meet.

Darren's passion for asset management and his love for teaching and working with students are so strong, that he is ready to take on the rocky path. He is convinced that the opportunities outweigh the challenges. Darren spends most of his time with graduate and undergraduate students, teaching them and reviewing their theses and papers. He has always wanted to share his knowledge and experience with the younger generation and to spread awareness about good physical asset management practice as much as he can. At the same time, he sees great opportunities to do research and to innovate within the field of physical asset management while he works on his PhD.

Darren's ultimate dream is to become a professor, but he realizes it will not be easy.

CHAPTER 1

PHYSICAL ASSET MANAGEMENT IN ORGANIZATIONS

"Assets are the livelihood of organizations. By definition an asset is anything that has potential or actual value to an organization." This is how Darren starts his conversation with Jerry Detunam, who has been waiting for him in the hotel lobby.

"Organizations have different types of assets that they own/use in their operations to generate income," Darren continues. "There are different types of assets in an organization:

- There are fixed assets, also known as non-current assets, which can be tangible, such as machinery, plant, and equipment, or intangible, such as brand, goodwill, etc.
- Current assets are inventory, cash, etc.
- Organizations also deal with financial assets, which can be in the form of current liabilities (short-term debts) or non-current liabilities (long-term loans, debentures, etc.).
- Another type of asset that organizations have to manage is people, which include their salaries, benefits, and expenses."

Jerry is attending a workshop at 9:30 in the hotel. So they have a little under an hour to chat. Darren had a quick breakfast on his way and managed to get to the hotel on time. He feels a little exhausted after his swimming workout and is relieved that they found themselves a quiet and comfortable corner in the lobby of the hotel to talk.

With his inseparable stainless-steel water bottle clad with the Ironman logo in his hand and slouching on the sofa, Darren continues: "So, the assets owned by an organization are accounted for in the balance sheet, also known as a 'statement of financial position.' The balance sheet is divided into two parts. On one side you have the assets, and on the other you have the liabilities and shareholders' equity. Technically, the two parts must balance each other out. Assets are what a company utilizes to operate its business, while its liabilities and equity are two sources that support these assets. In a balance sheet, there are two types of assets registered: current assets and non-current (fixed) assets."

"It looks like I am getting a financial crash course!" Jerry says with a smile.

"Yes, indeed. It is very important that you understand all the aspects of assets in organizations before you think about asset management," Darren adds. He explains: "Current assets have a lifespan of one year or less, meaning they can be converted easily into cash. Such asset classes include cash and cash equivalents, accounts receivable, and inventory. Inventory represents the raw materials, work-in-progress goods, and the company's finished goods. Fixed assets are assets that are not turned into cash easily, are not expected to be turned into cash within a year, and/or have a lifespan of more than a year. They can refer to tangible assets or physical assets such as property, plant, and equipment. Fixed assets also can be intangible assets such as goodwill, patents, or copyrights. While these assets are not physical in nature, they are often the resources that can make or break a company—the value of a brand name, for instance, should not be underestimated."

"Depreciation is calculated and deducted from most of these assets, which represents the economic cost of the asset over its useful life."

Darren sighs and continues: "Now let's talk a little about liabilities and shareholders' equity. Liabilities are the financial obligations an organization owes to outside parties. They can be in the form of debts and other non-debt financial obligations, as well as short-term or long-term borrowings such as debentures."

"How about employee salaries?" Jerry asks.

"Good question. Organizations invest in skilled human resources that are critical to operations. Human capital is regarded as the most important asset of an organization. On the balance sheet, salaries and wages that have yet to be paid are considered a liability. That is because it is money owed to a party outside of the organization. For salaries already paid, they can be treated as direct or indirect labor in the accounting system and even sometimes included in the asset section of the balance sheet if they're considered capital expenditures. The bottom line is that investment in human assets has to be accounted for in the financial reporting whether as a cost item or as an investment, which is another big debate in itself."

Jerry does not look quite content with the statement, but Darren continues with his explanation.

"Shareholders' equity," Darren says,"is the initial amount of money invested in a business. It is also referred to as the owner's residual claim after debts have been paid and is equal to an organization's total assets minus its total liabilities. Shareholders' equity represents the net or book value of a company."

Darren bends over toward Jerry and says in a measured tone: "To be financially sustainable, an organization needs to be able to maintain its financial capital, its fixed assets capital, and its human resources capital over the long term. If any of those go awry, it means the organization is not performing at its best."

"It is imperative for organizations to manage those assets diligently and realize maximum value in order to remain competitive and profitable. To effectively manage those assets, organizations are structured in such a way as to provide the required governance, leadership, and support to achieve their strategic objectives."

"Research that was conducted on over a hundred organizations globally shows that large municipalities manage an average 35 billion dollars of assets, while major companies manage an average of 100 billion dollars of assets."

Darren looks at Jerry with an air of confidence and resumes: "You walk into any organization and you can see that those assets are closely managed by a team of experienced and skilled staff. In many cases, you will even find that those teams are led by C-suite executives to provide better direction, planning, and execution, as well as compliance with industry regulations. Modern organizations have realized how important strong leadership, governance, and accountability are, because the executive team is a reflection of the organizational structure. It is the governing body that sets firm strategy, coordinates activities, and allocates resources across business units. Studies have shown that more and more organizations are increasing their number of corporate functional managers. This phenomenon is called 'functional centralization.' Functional centralization refers to the process to increase centralization of activities in organizations to realize synergies."

Darren starts making some gestures with his hands, a sign that he is getting excited by the conversation. Waving his hands around, he continues his explanation. He tells Jerry: "To exploit the synergies in organizations, activities from different business units have to be harmonized. An example of how corporate-level functional managers are used to capture synergies is Procter & Gamble's shift in 1989 toward a matrix organization. Functional senior vice presidents were engaged to manage functions across business units in order to promote 'the pooling of knowledge, transfer of best practices, elimination of redundancies, and standardization of activities.'" [Guadalupe, M, Li, H, Wulf, J, "Who Lives in the C-Suite? Organizational Structure and Division of Labor in Top Management," *Management Science* 60, 4(2013).]

"Since then, more and more organizations have increased the number of corporate-level functional managers to exploit potential synergies as part of their strategy. Corporate-level functional managers perform different activities that vary by function, such as the inherent primary (directly related to asset operations) and secondary (supporting operations) functions."

"Some of the common corporate-level functional roles found in organizations are:

- Chief Financial Officer (CFO)
- Chief Operating Officer (COO)
- Chief Human Resources Officer (CHRO)
- Chief Information Officer (CIO)
- Chief Marketing Officer (CMO)

All the above functional managers report to the Chief Executive Officer (CEO) or the Chief Administration Officer (CAO), who is responsible for the overall management of operations and resources of an organization and acts as the main point of communication between the board of directors or council and corporate operations."

Jerry looks at Darren with an inquisitive face, trying to figure out where this conversation is going to lead.

"Let us look at an example," Darren says with a change of tone. "A CMO is a relatively new role in many organizations, whose responsibilities include leading the company's marketing strategy, coordinating marketing activities, uniting and strengthening various departments' own marketing plans, directing global marketing efforts, and managing customer relationships. Historically, all marketing activities had been performed within the individual departments, which led to marketing campaigns all over the place, duplication of efforts, and inefficiencies. Similarly, the other corporate-level functional management roles are crucial to coordinating activities and leveraging synergies across different departments and the whole organization."

Darren starts describing the main duties of some of the C-suite positions:

- "Chief Financial Officer—Accountable for managing the processes for financial forecasting and budgets, and overseeing the preparation of financial plans and all financial reporting. Any organization will have a finance department to ensure that the finances of the organization are in order over the whole lifecycle of financial transactions. This applies to transactions ranging from accounts payable to accounts receivable, from operating expenditures to capital expenditures, through banking and investment.
- Chief Operating Officer—Accountable for the organization's business operations and ensures effective implementation of operational and financial procedures to deliver business goals. Develops strategies to manage raw materials, supplies, works in progress, and final inventories.
- Chief Human Resources Officer—Accountable for developing and executing human resource strategy in support of the overall business plan and strategic direction of the organization, specifically in the areas of succession planning, talent management, change management, and organizational performance.
- Chief Information Officer—Accountable for the implementation of suitable technology to streamline all internal operations and help optimize their strategic benefits. This includes designing and customizing techno-

logical systems and platforms to improve customer experience, data processing, and security."

"Why are we going through the C-suite roles and responsibilities, and what does this have to do with physical assets?" Jerry asks after listening to Darren with great patience.

With an unhappy look, Darren quickly replies, "What I am trying to highlight is: Do we have such a function for physical assets in modern organizations? Do we need such a corporate- or senior-level function to exploit synergies and coordinate activities related to physical assets?" He adds in a serious tone, "Physical assets are no different from any other assets owned by the organization, where someone has to be accountable for them."

Then he asks Jerry, "Can you imagine what percentage of an organization's assets is fixed assets?"

Darren looks at Jerry in anticipation of an answer. But Jerry just shrugs his shoulders and stares at Darren.

"Well, of the 35 billion dollars of assets from municipalities, an average of 28 billion dollars is fixed assets, representing property, plant, and equipment as taken from their financial statements. This is around 80%! For the major companies, of the average 100 billion dollars of assets they manage, 18% are fixed assets."

"This is quite a significant percentage, especially when you take into consideration how much value those fixed assets bring to the organizations," Darren adds.

Jerry seems to be now more relaxed as he starts to understand where the discussion is going.

"Historically," Darren continues, "management of physical assets was dispersed in the operations and maintenance functions. Over the years, the discipline of asset management has evolved from a normal routine set of activities of managing assets into a discipline that is now standardized by the International Organization for Standardization (ISO) and even regulated in some parts of the world. I have to admit that many organizations have come to realize the value that physical assets bring to their operations and are aware of the huge opportunities that exist to capture synergies and harmonize activities. Nowadays we see some proactive organizations already appointing vice-presidents, directors, and executives to be accountable for their physical assets and to lead their physical asset management strategy."

Jerry's job is actually a newly created position in Mouroum Inc. He left his job in maintenance and reliability to take on this new job, which is a move up for him in terms of pay as well as status. However, he did not realize that this is a completely different ballgame with a totally different approach. Jerry understands that he has already committed and he cannot afford to go back to his old position. With his current challenging family status—a newborn baby and a recently purchased new home—he badly needs to succeed in this job to make ends meet.

Jerry has been on the job for almost four months now, and he has been struggling—struggling to get an asset management program started, struggling to promote asset management as a priority for the department, and, most importantly, struggling to "fit in" in the organization.

Having known Jerry for quite some time, Darren knows what an intelligent and hardworking young man he is, and that he's always keen to learn. One of the first things Darren did was to mail Jerry a copy of his book so that he could read it before the meeting.

"Did you have a chance to read the book I sent you?" Darren asks.

"Yes, I read most of it, and I am, I think, on Chapter 7," Jerry replies.

"OK, that's good. If you recall, Chapter 2 explores the evolving journey of organizations to better understand and adjust their focus over time, and how they adapted to the ever-changing landscape and requirements. Obviously, the way physical assets are managed has also evolved over the years, driven by new expectations, regulations, demands, and other requirements."

Darren drinks a big sip of water from his bottle and lowers his voice. "Isn't it interesting," he says, "to understand how physical asset management itself has evolved as a discipline and what its roots were? How over time it has reached the level of the much talked about subject today? Look around you. There is not a day or week that you don't hear about a webinar, conference, or some kind of training or certification in physical asset management."

"The definition of asset management has come a long way, and, over time, there has been significant development across the globe in the field of physical asset management. It is beneficial for you to understand the evolution of asset management from a simple routine maintenance approach or a simple accounting/financial practice to combine and become a comprehensive field impacting whole organizations and in some areas even legislated (Figure 1.1)."

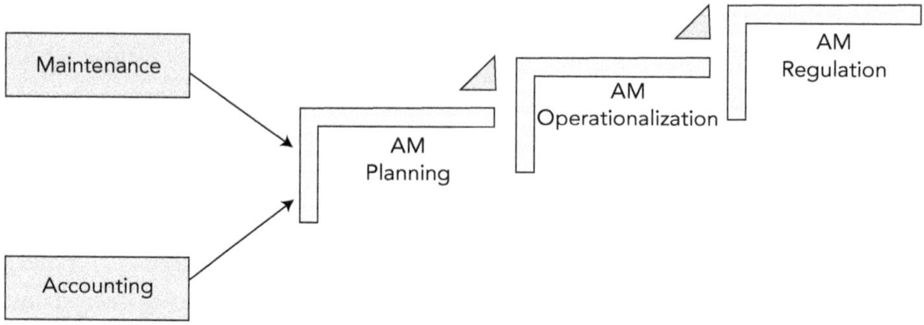

FIGURE 1.1 Evolution of asset management discipline.

"This is really interesting," Jerry replies. "I always thought that asset management was a new discipline and never bothered about its roots."

"Many say the coming together of the physical asset management discipline was triggered in the finance sector. Physical assets were engineered, installed, operated, and maintained since the beginning of time to support operations. On the other side, accounting and finance were reporting on physical assets in the balance sheet. At some point, organizations were required to provide more accurate reporting on tangible capital assets in their financial statements under the International Financial Reporting Standards (IFRS). With accurate reporting on physical assets in the fixed asset sub-ledger, we start seeing a disconnect between physical asset condition, consumption, and betterment and physical asset valuation and depreciation."

"I think this is the situation currently prevailing at Mouroum Inc., where there is poor alignment between accounting data and technical data in the field. I notice multiple inaccuracies and data gaps when I work with the finance team!" exclaims Jerry.

"This is common in many organizations," Darren continues. "This misalignment makes it very difficult for everyone to come up with a decent capital forecast and investment plan. Organizations need to improve planning for capital investment for their budgeting process. Private-sector companies need it to remain competitive and get the best returns on their investments, while the public sector needs to implement a rigorous process for asset management planning to forecast future expenditures. For example, municipalities in many countries are required to develop accurate asset management plans to be eligible for grant funding opportunities."

Jerry sits up and, with a happy voice, says, "This is actually one of my main mandates at Mouroum Inc.: to develop accurate asset management plans to meet the strategic objectives."

Darren nods his head to show agreement. He clears his voice and adds: "While considerable progress has been made by many organizations with respect to developing accurate asset management plans, there is still quite a lot to accomplish in terms of implementing the plans. We all know very well that 'A plan remains a plan unless it is put in action,' i.e., unless it is operationalized. In many organizations, the asset management plans are not being used in the operational and capital budgeting process, and, worse, they are not used to inform the decision-making process. By definition 'operationalizing' means setting something up (your asset management plan) so that it can be used, not only by a few functional areas, but across the whole organization."

Darren looks up Jerry and continues in a somewhat gloomy voice: "Unfortunately this is where we are nowadays in many organizations: stuck with executing the plans and operationalizing them organization-wide. This means getting all the functional areas dealing with physical assets in one way or another to fully understand, endorse, and implement the asset management plans in their respective areas. As you can imagine, this is not an easy feat, because we are now dealing with the whole organization: the governance, the different functional areas, the people, the culture, etc."

"I can see how difficult and challenging this will be," Jerry concurs. "Without any asset management plan to implement at Mouroum Inc., I am already getting pushback. Pushback in even understanding and embracing the physical asset management philosophy."

"I hear you," Darren says. "No wonder why, in some industries and countries, physical asset management is now regulated. There are a few good reasons for taking such a stance. The first one is to shift the focus and priorities of many organizations to get them to start thinking seriously about physical asset management. This, in return, will help to clearly identify what their infrastructure needs are and facilitate a more sustainable funding plan. Regulation to operationalize physical asset management planning also helps achieve greater standardization and consistency and reinforces the use of asset management plans as a tool to make better decisions to address physical assets needs and challenges."

"This is all great information," Jerry says after listening with attention. "But where do I start with an asset management program in my organization and how? One of the first things I noticed is that people tend to think that maintenance is asset management, and some people were even questioning my job. It is hard for me to explain to them, and they are not even ready to embrace the physical asset management philosophy."

"Well, the first thing you should start with is educating staff to spread awareness of physical asset management across the organization. There is so much documentation and various literature out there that is really informative. You can use those resources right away to help personnel from different functional areas understand the asset management philosophy. While doing this, I would also suggest you think about setting up a governance framework to launch and support your asset management initiative. Remember, alone you will not be able to meet the goals of an asset management philosophy," Darren explains.

"What does that entail?" Jerry questions.

Darren chuckles and replies to Jerry: "That can be a lengthy conversation, and, unfortunately, I have to go. Can we continue our discussion sometime next week?"

"Absolutely," Jerry responds. "You tell me when you're available. In the meantime, I will search the training/educational materials as well as governance frameworks. I agree there are many resources available on the internet, but I will need your guidance on how to use them."

"That's a good idea. But I would caution you that 'how' to use the information is not enough. You also need to understand 'why' to use those resources and 'when.' Picture it as a recipe with all the ingredients. There is a particular order and timing at which each and every ingredient is added. There is a specific reason for this."

"Think about it, and I'll see you next week," he adds with a smile.

Darren gets into his car, takes a deep breath, and starts checking his phone for any missed calls, especially from Debbie. There are none. Darren and Debbie have been happily married for twenty-seven years. As in any couple's life, there have been ups and downs. Most of the time the sources of conflicts have been due to Darren's busy schedule at work and his other activities. For the last few years, Darren's involvement in the triathlon and, just recently, his decision to take early retirement to pursue his academic dream have not eased the situation. He knows he has to give Debbie more of his time. Darren puts on one of his favorite tunes and drives away. He is feeling so tired and just wants to get home for some rest.

CHAPTER 2

THE GOVERNANCE FRAMEWORK

It is a beautiful Wednesday evening, and Darren drives into the parking lot of the coffee shop. He has just finished a busy day at work reviewing a few draft papers that his students have been working on. Darren knows that Jerry is desperate to talk to him. It's been two weeks now since they last talked. He twice postponed Jerry's meeting requests, and he thinks that it won't be good to postpone a third time.

As he enters the coffee shop, he sees Jerry sitting at a table in the far corner next to a large window. He must have seen him when he drove in. Actually, maybe he did not notice him, as he seems to be deeply immersed in reading something on his laptop, and he has a stack of papers next to him.

"I see that you have been killing a few trees. Think about sustainability, my friend," Darren says as he stands right in front of Jerry's table. "So what have you found out, and how is Mouroum Inc. coping with asset management and its governance framework?" Darren asks.

Darren actually startled Jerry, as he suddenly looks up almost toppling his laptop.

"Good evening, sir," Jerry greets him. Jerry has always been a very respectful and polite gentleman, and that is maybe why Darren likes him so much. Darren finds it very difficult to turn him down and not respond to his request for help, even though he is himself running around like a headless chicken.

Darren takes a seat at the table and puts his bag on the vacant chair next to him. Jerry orders two coffees and then starts providing Darren with an update on his progress.

"First of all," he starts, "you were right in saying that there is a ton of information out there available on the internet. I can see that there is a broad range of asset management approaches and methodologies, varying in complexity and sophistication, that are in use in different industries and organizations."

"Yes," Darren nods. "However, while there are existing pockets of best practice, it is recognized that a more consistent approach to the management of physical assets will yield significant benefits. Therefore, organizations are recommended to adopt a corporate approach to the physical asset management framework spanning across the whole organization."

"What do you mean by that?" Jerry asked. "I mean by a 'corporate approach'?"

"What I mean," Darren starts to explain, "is that the approach must be blessed, endorsed, and championed from the top to the whole organization. It does not matter the size and complexity of the organization, but if organizations recognize and agree for the need of an asset management program, then the commitment has to come from the top down. No matter if it is called corporate or not, but the commitment and buy-in must be there. Otherwise, it will be another of those initiatives that fizzle out after some time without any accomplishments. The ultimate aim is to develop and implement consistent asset management approaches across all functions and departments of the organization."

"This is exactly the challenge I am facing in my job at this moment. I am having a hard time getting our asset management program started, and there is no buy-in," Jerry exclaims.

Darren drinks a sip of his hot coffee and says: "I understand what you are going through, and I would advise you take a well-calculated approach as you are about to take your organization on a long transformational journey. You'll definitely want to get off to a good start."

Step 1: Develop an Asset Management Governance Framework

"The first step I would recommend to you is to start developing an asset management governance framework. Too many times we see organizations diving into projects or programs focusing on short-term gains only. They do not bother to have a framework or structure in place to sustain the long-term outcomes of what was started. The result is that either the project or program does not deliver expected outcomes, or it just dwindles over time and loses momentum or interest. If you want to build a long-lasting physical asset management system in your organization, a governance strategy is the foundation piece that I recommend you start with."

Jerry looks excited, and he starts taking notes on his laptop as Darren speaks.

"The governance framework will be a key component of your physical asset management system that will ensure consistency in asset management strategies across the organization via the adoption of policies, practices, procedures, templates, and standard tools," Darren explains. "A governance structure provides effectiveness and efficiency by constantly sharing reliable and relevant information in the asset management decision-making process through a standard approach and formal definition of asset management roles and responsibilities. The governance framework takes into account both the short-term delivery and the longer-term asset management system sustainability. The key objectives for an asset management governance framework can best be described as follows:

- To provide an organizational design that will strategize implementation of a physical asset management system
- To provide an organizational design that operationalizes integrated and comprehensive asset management practices across all functional areas
- To provide clear strategic roles and responsibilities for staff involved to enable and facilitate the transformation process as required
- To ensure consistency in asset management concepts and practices across the whole organization at the strategic, tactical, and operational levels"

"How do I get started with developing such a framework?" Jerry calmly asks. "I think this is what is missing in our organization, where I am constantly struggling to pull everyone together to dedicate time to work on the asset management program."

"In Chapter 12 of my book, I talk about 'ownership' and the 'ideal organizational structure' to lead the implementation of and maintain a physical asset management system. It also elaborates on three different governance models and does mention the fact that a lot will have to do with the context of your organization. First of all, you will have to understand the external and internal context of your organization: What are the social, cultural, political, legal, financial, economic, and environmental factors having an impact on the strategic objectives of the organization? What are the organizational culture, values, size, complexity, policies, strategies, and capabilities driving the needs for a physical asset management system?"

Darren watches Jerry type on his laptop as he takes a swig of coffee.

"Then you will need to identify the stakeholders relevant to your organizational physical asset management system and understand their needs and expectations," Darren says. "Make sure you consider both internal stakeholders such as the shareholders, owners, different departments, functional areas, and employees, in general, as well as external stakeholders such as customers, users, suppliers, contractors, regulatory authorities, and other agencies."

Darren takes a pause and adds, "Mind you, the lists I mentioned to you are just examples. I would suggest you have a few sessions with your team at different levels in your organization to brainstorm and come up with a list of external and internal factors that would define your organizational context, as well as a list of applicable external and internal stakeholders with their needs and expectations."

"This looks like quite a lot of work!" Jerry exclaims. "It will take quite some time to get this done."

"I couldn't agree more," Darren continues. "I can guarantee you, though, that it will be time and effort well spent. At the end of this exercise you will get a pretty good idea of the context of your organzation, and, on top of that, you will have started engaging key stakeholders in the process. You will be able to recommend a governance structure that will fit your context. The governance model chosen will have to provide stakeholders with the necessary autonomy and flexibility to interpret, adopt, and implement physical asset management approaches and practices into their existing activities."

Darren adjusts himself in his chair before cautioning, "Always keep in mind that the purpose of a governance model is to ensure that:

- There is an organization-wide common and standard approach to physical asset management.
- Roles and responsibilities of stakeholders are clearly defined and communicated.

In short, the governance model should embody the strategies, policies, and structures that provide the mandate for the strategic/tactical framework. This, in turn, structures the necessary procedures and practices to guide appropriate decisions and actions. Hence, you would want a governance structure that will engage key stakeholders at different levels in the organization as well as having a championing body that will endorse and support the physical asset management system."

"Governance models can be different for different organizations based on their size, complexity, and context," Darren adds as he reaches into his bag and pulls out a piece of paper and a pencil.

Jerry watches in silence as Darren starts drawing a few rectangles. After a few minutes Darren hands Jerry the paper (Figure 2.1).

"Here you go," Darren says. "This is a governance model that I used in my previous job and worked pretty well. You can start playing around with it to adapt it to your organizational context."

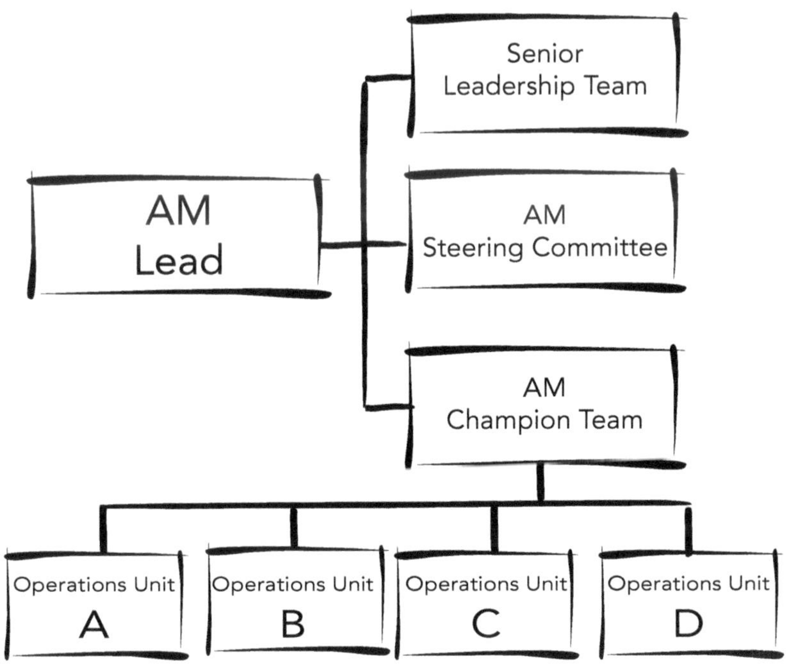

FIGURE 2.1 Centralized asset management governance model.

"Thank you so much, sir!" Jerry grabs the paper in delight. "This will definitely help me getting started."

Right away Darren warns Jerry, "Mind you, Jerry, this is only an example of a governance model: a centralized corporate asset management office, and an asset management steering committee, and a decentralized asset management network team. The steering committee is responsible for giving directions and supporting the asset management initiatives, while the asset management office is responsible for developing and implementing an asset management system. The network team is responsible for rolling out and operationalizing the asset management initiatives."

Darren leans over the table and clasps his hands as he says: "Your organization may require a completely different model that will suit your organizational context and ensure consistency and sustainability of your asset management system. It will also need to deliver expected results while maintaining a certain autonomy and flexibility for asset owners and operators to interpret and implement the asset management processes and procedures."

Jerry looks at the sketch for a couple of minutes and then looks up with a smile. This indicates to Darren that he agrees and is on board with the concept.

"The next thing I suggest you focus on," Darren continues,"is education and awareness."

Jerry looks at him a little bemused.

"Yes, you heard it right," Darren continues. "At this stage, it is of utmost importance to get to know your stakeholders and to engage them early on to set the tone for a collaborative and supportive environment across all the functional areas."

Step 2: Develop an Asset Management Framework

"To do that, you need to reach out to the different stakeholders not empty-handed but with something to help them understand what asset management is, what you are planning to do, and where they fit in. At the end they should see how the whole process would benefit them in their day-to-day job. It is basically answering the obvious question, 'What's in it for me?'"

"This seems to be a reasonable thing to do—especially in my situation," Jerry says. "I often get asked questions such as 'What is asset management?' or 'Why do asset management?' I do my best to explain, but I guess I do not do a good enough job as staff still look confused," Jerry continues.

"Don't be too hard on yourself," Darren tries to encourage Jerry. "It's not you; it's just the way things have evolved and, with it, lots of misconceptions about asset management."

Darren goes on, "In this situation an asset management framework does address those misconceptions. Imagine you have a puzzle with different individuals looking at the different pieces of the puzzle on their own. You would agree that it won't make any sense whatsoever even if you take all the pieces of the puzzle and place it on a table. It is only when you have assembled all the pieces together that you will see the real picture of what the puzzle represents."

Darren mimics building a puzzle on the table as he continues. "That's exactly what an asset management framework will do," he says. "It will assemble all the pieces from different functional areas and put them together to show how they are connected to each other and, most importantly, what asset management is all about."

"Oh, I see," Jerry replies. "Then if I show this to the different departments, they will be able to see where their pieces of the puzzle fit in the whole scheme of things."

"True, and they will also be able to understand how their work relates to other work in the organization and how this can impact the overall objectives. There are many asset management framework templates out there, and it is easy for you to review them and create your own based on your organizational context. I have seen simple ones as well as some complex ones."

As Darren continues to talk, he pulls the paper with the governance model sketch from Jerry, turns it over, and starts working on another sketch.

Darren is a very visual person and thinker. He likes to draw concepts and images. It is not a surprise that one of his hidden talents is actually sketching and drawing. This time he could not stop himself from having some fun with quickly drafting an asset management framework for Jerry.

Darren is sure Jerry will appreciate the sketch as he watches him do his artwork with a smile.

After a few minutes of drawing and shading, Darren hands the paper over to Jerry once again (Figure 2.2).

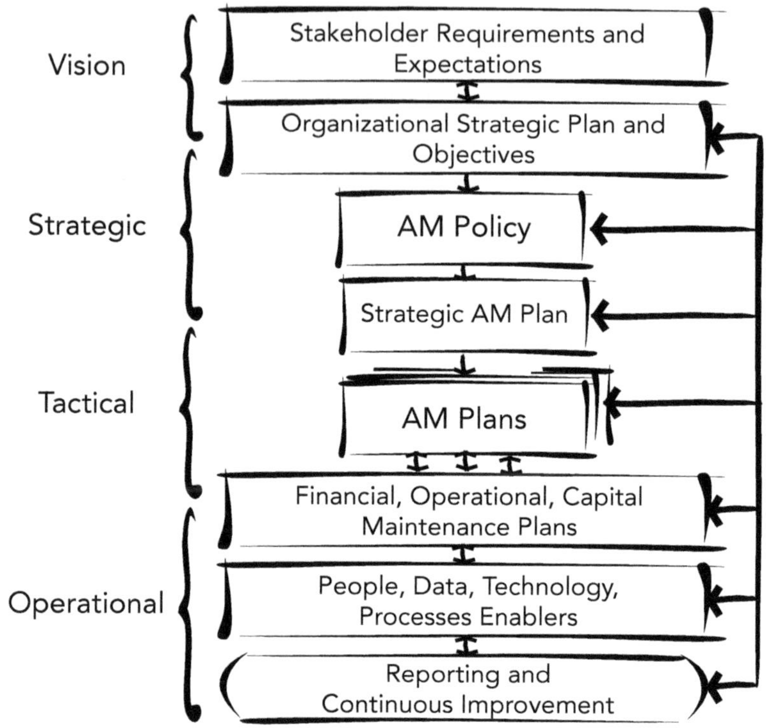

FIGURE 2.2 Asset management framework.

Jerry grabs it and starts studying it closely as he tries to follow and make sense of the lines and boxes.

As Jerry studies the sketch, Darren explains to him what he should look for when he develops the asset management framework: "The asset management framework needs to consider the broader view of the organization, meaning it has to consider the different levels in the organization, from strategic, to tactical, to operational. It also needs to show the line of sight from a whole lifecycle physical asset management perspective across all the different levels and demonstrate how everything fits together."

"Can you please provide some examples of what needs to considered in the framework?" Jerry asks.

"Key elements of the framework are:

- The stakeholder's requirements and expectations
- The organizational plan and objectives
- The different strategies needed for the asset management strategy, namely, lifecycle management, level of service, risk, and demand management strategies
- The tactical planning required for different asset classes: financial, operational, capital, and maintenance planning
- The actual execution of the lifecycle delivery activities over the entire life of physical assets to deliver required performance
- The enablers to support execution such as technology, people, data, and processes
- Performance monitoring and reporting
- Continuous improvement"

Darren lets out a deep sigh and continues: "Remember, asset management is intended to provide for the sound stewardship of physical assets, meet customer service commitments, and manage assets in the most cost-effective and efficient manner. This essence is clearly captured in the framework."

"Once you have created your own asset management framework, I suggest you meet with your stakeholders, give them copies, and explain to them what the framework represents. Clearly show them where they fit in the framework, how their work relates to others, and how their work is important to achieve the goals of the organization."

"Finally, you can point to them what your role is in the whole framework and how you fit in."

Jerry looks at Darren as if he does not quite understand how he fits in the asset management framework. He carefully folds the sketch once and places it in one of his folders lying on the table.

Darren glances at his watch to check the time. He figures he still has around thirty minutes before he should really head out, as Debbie will be waiting for him for dinner.

"There is another key document that I would like you to think about or even start working on," he continues.

Jerry stares at Darren with a surprised look on his face. He must be feeling really overwhelmed with what Darren has been throwing at him so far. However, he does not say a word, and Darren keeps going, cautioning, "Next it is very important that you have an asset management policy in place."

Step 3: Develop an Asset Management Policy

"Jerry, do you have a health and safety policy in your organization?" Darren asks.

"Yes," Jerry replies.

"How about an environmental policy, do you have one?" he again asks.

"Yes, I think we do have . . . not so sure, though," Jerry responds.

"I am pretty sure you do have, especially with the business you are in. You must also have many other policies at different levels in your organization, such as an accounting policy, personnel policy, procurement policy, and many others. Therefore, the question is why we do not have one for asset management yet?" Darren replies. "We have policies to protect and manage other types of assets, so why not for our physical assets?"

Then Darren reminds Jerry, "Last time I did mention to you the significant value of fixed assets that organizations end up owning and managing. These same fixed assets can be the source of many health and safety, environmental, or other consequences. The irony is that not having a commitment to manage physical assets in the form of a formal asset management policy can directly impact the goals of the other policies."

Darren takes a pause and asks, "Isn't this weird? Physical assets and their contribution have always been taken for granted or gone unnoticed in organizations—especially at the boardroom level."

He pauses again to collect his thoughts and then says: "This has to change, and, actually, I should correct myself . . . things are already changing these days. Many organizations have started to recognize that there is a gap in their commitment to responsibly manage physical assets. Organizations in general do an outstanding job in developing their strategic plan to outline their vision, mission, and strategic goals and objectives. From there they develop their long-term and short-term financial, marketing, and operating plans. The physical asset management portion is dealt with to some degree in the operating plan. However, there exists a considerable gap in this area, and this is what many industries are trying to address these days."

Then Darren starts on a relatively long explanation of what an asset management policy is all about: "A *corporate* asset management policy helps formulate the commitment to look at the strategic aspect of managing physical assets. The policy outlines the multiple approaches and requirements the organization endorses and commits to for better asset management planning. It spells out the underlying guiding principles to follow when developing and deploying asset management practices to meet legislative requirements and the organizational strategic objectives. In order to ensure that the asset management system is compatible with the rest of the organization, it has to be clearly specified and defined through a formal policy statement by the organization as a whole. This action is essential so that the organization places the asset management system in context within its overall range of strategic objectives and goals. This is necessary both to give the development of the system visible and distinct credibility within the organization, and to ensure that the goals of the system and the goals of the organization are correctly aligned and calibrated. The establishment of a policy is also the first step in a longer process of developing individual and specific asset management objectives for different functional areas of the organization."

During all this time Jerry has been listening carefully while trying to capture as much as he can in his notes. Clearly he is trying his best to absorb all the information.

Darren takes notice of that and understands that maybe it is a bit too much for the young man. He looks at his watch and notices that they have been meeting in the coffee shop for more than two hours already.

Darren gets up and says, "I think that's enough for today. I think you have enough to work on for the next two or three weeks. Don't worry about the rest; just be patient and focus on engaging and educating your stakeholders. Next time we meet, you will tell me of your progress and how things went."

Jerry does not say anything. He just shakes Darren's hand and gives him a little, almost hesitant smile.

"All the best," Darren says as he walks toward the exit.

CHAPTER 3

THE STRATEGIC ASSET MANAGEMENT PLAN

For the last couple of weeks, Darren has been quite busy. On one side, he has been busy teaching the three courses he has ongoing, and, on the other side, he has been catching up with reviewing some papers his students have drafted.

At home, also, things have been quite busy for him. He tries to spend some good quality time with Debbie as much as possible, while at the same time training hard for long hours to prepare for his next triathlon. With his medium height and slim build, Darren is very fit and does not look his age at all. If it were not for his salt and pepper hair and shortly trimmed beard, he could easily be mistaken for someone in his mid-forties. He is often asked how he manages to keep up with all his activities. To tell the truth, Darren himself doesn't have a clue at all. He just plans what he needs to do over the year and months, looks at the upcoming week, and takes it day by day. Two of his best-kept secrets are "stick to the plan" and "be consistent." Debbie plays an instrumental role in keeping him focused and to the task, as she sometimes surreptitiously adds her dose of pressure. She knows very well how and when to do it so as to keep Darren in line. Darren is fine with Debbie's approach and actually enjoys it, as it allows him to practice his preferred activities while staying focused and aligned. This is something that Darren has witnessed many organizations struggle with: *performing all the required tasks in coordination while maintaining alignment to deliver on strategic objectives.*

Training for Ironman races has taught Darren a lot about the importance of having a vision, setting goals, developing a strategy and plan, and executing the plan according to the strategy. Also important for him is constantly measuring his performance to continuously improve while at the same time doing as much research as possible. Darren is always ready to change and to adapt while leveraging new tools and techniques to achieve the right balance, discipline, and consistency required for success.

Step 4: Strategic Asset Management Planning

It is a beautiful Sunday, a little chilly outside. Debbie and Darren just came back from their morning swimming session. They have been doing this routine for years and enjoy waking up early and having their workout. Debbie is a petite woman but very physically fit. Typical Sunday breakfast consists of lemon juice, half an avocado with honey, toasted brown bread with a boiled egg and cheese, half a banana, and a big cup of tea. They enjoy chatting and watching the news as they have breakfast. After breakfast, it is time for some household chores.

By around eleven o'clock Darren makes himself comfortable in the family room, switches on the TV, and selects a movie. He then fires up his laptop and starts catching up on some of the draft papers from his students. One of them—written by Neel, a brilliant student in his second year—caught his attention. The title of the paper is "Why Physical Asset Management Plans Fail," and it talks about some of the reasons why organizations are not successful in implementing their asset management plans and fail to realize the best value from their portfolio of physical assets.

One of the key factors cited is a lack of strategy in organizations to operationalize their asset management plans. It talks about the importance of having an asset management strategy and asserts that organizations are in such a hurry to put together their asset management plans, they forget to consider the strategies needed to implement them. In his paper, Neel goes to great lengths to explain what an asset management strategy is and what should be included in it. He references ISO 55002's definition of a strategic asset management plan:

> Documented information that specifies how organizational objectives are to be converted into asset management objectives, the approach for developing asset management plans, and the role of the asset management system in supporting achievement of the asset management objectives.

Darren continues reading the paper, as it is really very well written and makes very much sense in the context of how organizations should be managing physical assets. Darren believes that a few sections like the ones below will be very beneficial to share with Jerry the next time they meet. The sections explain the purpose of a strategic asset management plan (SAMP):

> The SAMP sets out a strategic framework that guides future asset-related initiatives to support the organization's strategic plan. The plan documents and specifies how the organizational objectives are to be converted into asset management objectives, the approach to develop asset management plans, and the strategies to build and support an asset management system. It helps organizations to set up a framework to be able to make fact-based decisions regarding the building, operating, maintaining, renewing, replacing, and disposing of physical assets to maximize benefits, manage risk, and provide satisfactory levels of service.

In the draft paper, Neel goes on to explain why organizations sometimes fail to implement their physical asset management plan:

> Organizations spend time and money preparing fancy binders of asset management plans which end up sitting on shelves collecting dust. The plans are sometimes not executed because of a lack of strategy.

He also explains that it is important to differentiate between strategy and plan:

> A business is managed through the development and execution of strategies and plans. An organization may have a plan to roll out a new product. Research, design, manufacturing, marketing, and sales will all happen at specific times according to the plan. The new product will continue to go through the same process, but behind the plan there is a strategy for the new product to be innovative, with an improved design, manufactured with the latest technology, and targeting a certain market to ensure it becomes a success. Why is the same approach not applied to managing physical assets?

Neel clarifies the confusion that exists between the terms "strategy" and "plan," which are often mistakenly used interchangeably. He explains:

> A plan is usually a list of steps taken to accomplish a goal and tackles questions like what, when, who, etc. A plan is always a good thing to have in place to be successful, but it should not be the first step in addressing a task (Figure 3.1).

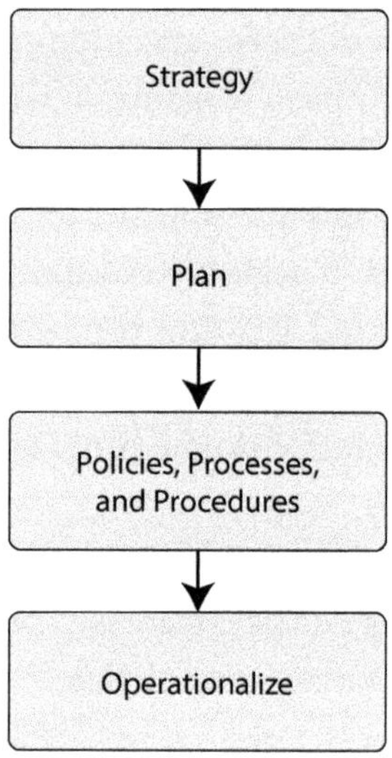

FIGURE 3.1 Strategy and plan.

> A strategy, on the other hand, is bigger than a plan. Strategy tackles the question of why. It has a large scope and looks at the end result as well as the many paths to the desired outcome. Strategy speaks to the reasons why, whereas the plan is focused on how.
>
> A strategy is an overarching framework, a high-level, big-picture, and long-term approach that spans across the whole organization. Given that physical asset management does impact many functional areas in an organization, it makes total sense that a strategy in some form is critical and should come before any physical asset management plan. A well-crafted and well-communicated strategy will achieve consistency and promote collaboration, two key aspects for successful implementation of physical asset management plans.

While reading the paper, Darren starts to realize that this is absolutely true when he looks back at how he actually approached his first Ironman race. He remembers setting the goal and developing a three-year strategy to get to the fitness level required to participate in his first Ironman race. The three-year strategy was broken down into strategies for different aspects of the race, e.g., strategies for swimming, biking, running, nutrition, and equipment and gears required. For example, one of the strategies he had was to focus on "try-a-tri" distance in the first year, a sprint distance in the second year, and an Olympic/long distance in the third year. To execute the overarching strategy and meet his goals, Darren remembers developing training plans for the different strategies with realistic goals to achieve over the years. He realizes that if he hadn't had an overall strategy in place, but instead started working with only training plans, he may have found himself all over the place with his fitness and performance. Maybe he could have been successful with some luck . . . But it is never a good idea to leave things to luck.

Neel talks further about the lack of understanding of the importance for organizations to have a strategic asset management plan. He compares how other functional areas have a strategic approach as well as a plan in place to deliver results, while the physical asset management field seems to be somewhat lagging behind. Very often you will hear about corporate strategy, finance strategy, human resources strategy, or even IT strategy. What about a physical asset management strategy? he argues.

Neel elaborates on what a strategic asset management plan should consist of and what role it plays. First of all, he highlights one caveat that practitioners should be well aware of: the difference between an *asset management strategy* and a *strategic asset management plan*.

Neel further clarifies that an asset management strategy is the strategic approach that an organization takes to realize maximum value from its physical assets. It ensures that alignment across all functional areas allows coordination in delivering asset management activities and promotes collaboration during decision-making processes to achieve a common goal.

On the other hand, Neel explains, a strategic asset management plan consists of applying strategic actions to deliver the asset management objectives derived from the organizational strategic objectives. It is a high-level, long-term action plan that includes objectives for managing the organization's physical assets.

The draft paper goes on to describe the process of developing a strategic asset management plan, arguing that, in order to develop a plan, it is necessary to apply the asset management strategy to "something with a goal." The "something with a goal" is the physical asset stock and the value the organization wants to derive from it. This means that, in advance, the organization needs to know the physical assets it owns, their current conditions and risks, the estimated service lives, and the estimated rehabilitation and replacement costs required to maintain them in good order to deliver the required levels of service at minimum risk.

In addition to that, organizations need to know what value means to the external and internal stakeholders. This is important so as to be able to determine the organization's physical asset management objectives. Usually for proper alignment, the organizational strategic objectives are converted into physical asset management objectives which would relate to effectiveness of the physical asset management system as well as the performance of physical asset classes/systems. Neel finds that this is an area where significant work needs to be done and there is general confusion about what physical asset management objectives should be. He cites two examples—one for a physical asset management system and the other for a physical asset system:

- To apply an optimized decision-making process for future capital investment
- To achieve 90% reliability in the processing plant

Neel argues that the issue is that organizations do not do a good enough job at documenting physical asset management objectives. The main reason is that physical asset management is not seen as a strategic initiative, and as a result their objectives are either buried in other functional areas or are sometimes neglected outright.

To build a sound strategic asset management plan, Neel reiterates that organizations need to look at the following two key components:

1. The physical asset management strategic approach the organization needs to take to ensure alignment
2. What the organization owns in terms of physical assets and what it will take to keep them in a state of good repair

Neel's paper stresses the importance of having both components developed in sequence, as one feeds and informs the other to produce a solid SAMP document.

"What is this again that you have bought?" Debbie looks at Darren while holding a white package in her hands.

Darren grabs it from her and starts ripping the packaging apart to open it. Debbie stands there, both hands on her waist watching him. Darren takes the item out of the box and holds it in his hand turning it around a few times as if he is inspecting it. He has bought a new nutrition pouch for his tri-bike to replace his older pouch, which he does not like any more. This new pouch fits on the frame of the bike in front of him, is a little bit roomier, with compartments, and, on top of that, is more aerodynamic.

"You always have money to spend on these things!" Debbie exclaims with her hands still on her waist. "For the last two years I have asked you to redo the laundry room with proper cabinets and to install recessed lighting outside the house. I am still waiting for these things. You keep telling me we have to save money for that, and yet I see that you are spending money left and right on your triathlon things."

Debbie is right. For the past few years, Darren has spent a small fortune on things for the triathlon—his bikes, wetsuits, trisuits, and shoes; his swim, bike, and run apparel; and different gears and accessories, plus the cost of training programs, registration fees for races, and the travel expenses to attend those races. All these little things add up over time and can become a significant separate budget.

As a result, he has a ton of triathlon junk in his basement, in the garage, and in his closet that needs to be sorted out. Maybe Darren should start with doing an inventory of what he has, assess the condition things are in, select the ones that bring value to him to keep, and identify those that can be disposed of.

Darren shakes his head and decides to deal with his triathlon asset management stuff later. For now, the main thing is to escape Debbie's wrath and let her cool down. Darren gives Debbie a big smile, and, without a word, he makes his way in the basement where he has his bike set up on his trainer. Within minutes he installs the nutrition pouch on the bike frame and jumps on his bike to start riding along with a group of cyclists online. He is riding on one of his favorite circuits, the Volcano Climb. He gets into an aero position and cruises downhill, with the volcano in the background spitting large streams of red-hot lava. He loves the beautiful scenery as he rides hard with drops of sweat trickling from his forehead.

"Wake up! Wake up!" Darren hears a voice in the background calling. He opens his eyes and sees Debbie standing over him and smiling. Darren is confused and does not understand where he is. The last he knew, he was just on his bike in the middle of a ride. After a few seconds he realizes that he had dozed off while reading Neel's paper. He looks at the clock and notices that it is half past noon.

"We won't have time for lunch," Debbie says. "Darren, you have to be ready quick; otherwise we will miss the start of the movie."

He actually forgot that they are supposed to go out today to the movies. The show they are targeting starts at quarter to two. He better get ready quickly, and grab a quick bite, so they can head out to the movie theater before it's too late.

One thing for sure is that during the show Darren will not fall asleep, because he has already had his few minutes' nap coupled with a very nice dream on his tri-bike. He jumps up and gives Debbie another one of his killer smiles and disappears upstairs to get ready.

The movie is, in fact, quite interesting, and both Debbie and Darren like it a lot. It was the type of thriller suspense movie that keeps you focused and tense for the whole duration. After the movie, they decide to go for a walk in the nearby park talking about the boys and how much they miss them.

Rémi and Lucas are both in their twenties now and not living with them anymore. Rémi has moved out since he got married last year. He lives about thirty minutes' drive from them and is enjoying his married life. Lucas is in his last year in university and lives on campus. He comes home two or three times a year. With Darren's busy schedule, one can imagine how lonely Debbie must be sometimes (or very often). Darren, however, does his best to be with her, and she understands him even though sometimes she can get upset.

After about half an hour of strolling, they decide to drive back home before it starts to get dark and chilly. Overall, it is a very relaxing time for both Debbie and Darren. As usual at this point, Darren is already starting to think about the next busy and exhausting week ahead. Lately he has been quite worried about the progress he is making to get his PhD and how long it might take. He is especially concerned about how long he and Debbie can sustain themselves on the salary of an adjunct nontenured professor. Surprisingly, a few times, Darren thought of quitting, which is very unlike him.

CHAPTER 4

PHYSICAL ASSET MANAGEMENT STRATEGY

"What should I wear for dinner tonight?" Darren asks Debbie. Debbie is used to hearing this question, especially when they are getting ready to go out. Yes, tonight they are invited for dinner at Jerry's place. Jerry has been trying to meet up with Darren for the past few weeks, but they have not been able to connect due to conflicting schedules. Then last week Jerry sent a text to invite Darren and Debbie over for dinner at his place. It was great timing, as it is a good opportunity to reconnect and continue their conversation from where they left off four weeks ago.

While Darren waits for Debbie to get ready, he gathers a few documents that he is planning to bring over to Jerry's house to continue their discussion on his challenges in asset management. He has especially taken Neel's paper, with some notes that he made, to remind Jerry of a few key items. He only read half of the draft paper and did not have a chance to go through the rest. Anyway, deep inside he knows it will be of great help to Jerry.

The drive to Jerry's house is around forty minutes. He lives in a new subdivision where he bought his house last year and moved in a few months ago. It is a beautiful townhouse in a very quiet neighborhood.

As soon as they ring the doorbell, Jerry opens the door as if he has been waiting for them right behind the door. They are welcomed inside with a large smile and invited to take a seat in the living room, which is immaculately decorated; light background music is playing. Debbie and Darren are greeted by Rita, Jerry's wife, who right away makes them feel at home with her lovely and charming personality.

Jerry serves the guests some red wine, and they start chatting. Naturally after a few minutes, the conversation between Darren and Jerry turns toward their most passionate topic, asset management.

"Jerry," Darren asks. "So how have things been turning out for you since we last met? I think it was in the coffee shop last month. Have you had any success with some of the ideas we talked about around education and awareness?"

Jerry's rubs his hands together, and the large grin on his face tells it all.

"Oh, it has been going very well, much better than I was expecting," Jerry replies delightedly.

"We have a governance model in place, and, last week, we just had our first steering committee meeting. It feels so good, you know, to have senior management at the table, learning about asset management and supporting the program. Suddenly, for the first time at Mouroum Inc., I felt I was not alone anymore in all this."

"Come on, Jerry," Darren jokingly challenges Jerry, "You were never alone in this—you always had me around, didn't you?"

Jerry blushes and feels bad for a moment. He realizes that it is actually true. He could always rely on Darren whenever required, and he was never alone.

"True," Jerry adds remorsefully. "I should not have said that. I was never alone in this. What I meant was in my work environment I was alone fighting through the day-to-day challenges. But I know that I always have you when needed."

"That's fine," Darren quickly adds. "I was just pulling your leg. Come on, tell me the rest."

Jerry continues that now he has some action items in place after the first steering committee meeting. He presented a draft AM framework that the committee liked a lot. He now needs to refine it and send it around for approval. He also talked about the need for an AM policy and was able to show the committee members some of the AM policies which are in use in other organizations. The committee sees the importance of such a policy and gave Jerry the marching orders to draft one for Mouroum Inc.

"I see that you are off to a great start Jerry," Darren says with joy.

"Yes, I am so excited, and I want to thank you for the great advice. Now I want to know what to do next. That is why I have been bugging you to meet with me."

"Jerry, you may regret what you are asking for. I have a lot already lined up for you."

"How about we discuss all that after we eat," Darren continues, having noticed the awkward look from Debbie.

Jerry and Darren change the topic of conversation, and they start talking about real estate and the new area where Jerry lives. This brings Debbie and Rita into the conversation, which makes it more interesting for everyone. They all have another round of red wine before dinner.

"I love this delicious food!" Darren exclaims. "Rita, you are a true cordon bleu!" he adds looking at Rita, who blushes and smiles at him.

Rita has prepared braised lamb shanks well cooked with onions, carrots, and garlic. She also made roasted spicy eggplant, all served with white rice and salad.

"Thank you, Mr. Darren," she politely replies.

Debbie gently taps Rita on her shoulder and murmurs something to her. It looks like Debbie also likes the food and is praising Rita.

They finish dinner and start working on dessert—a small bowl of bread pudding mixed with almonds and dried raisins. It is equally delicious!

While enjoying the dessert, Darren starts talking to Jerry about Neel's paper and explains to him the importance of developing a strategic asset management plan as the next step. Darren takes Neel's paper, which contains his notes and hands it to Jerry. As Jerry starts to look through the document, Darren continues to speak. First, he explains to Jerry, he has to get a good handle on Mouroum's total physical asset inventory and understand the condition of the physical assets.

"You may already have all this information available in your organization," Darren says. "The information may just be dispersed and is in different formats. Somebody needs to gather, clean, and arrange it in a standard structure. You can do it in-house if you have the time and resources, or you can use external help. Either way, I caution you not to try to get a perfect and accurate database. Don't worry if some information is missing or is not up-to-date. You can fix that later. The most important thing is that you have something to start working with even if not perfect."

"Excuse me," Jerry interjects, politely as usual. "I understand the asset management framework, the policy, and the inventory. How does the SAMP fit into all this?"

"Excellent question, Jerry!" Darren exclaims. "I was actually expecting that type of question from you. Let me think how to better explain it to you."

After a few seconds of thought, he begins by asking, "Do you have your organization's asset management framework?"

Jerry goes into his den and after a few minutes he returns with a binder from which he pulls out a page.

Darren places the sheet on the dining table as Jerry comes over to stand next to his mentor.

"Look here," Darren starts to explain while pointing to the top of the asset management framework above the area representing the strategic asset management plan. "The corporate asset management policy is like the glue that brings together and aligns the organizational strategic objectives with your strategic asset management plan. Picture for a moment that the SAMP does not exist. You will have a void between the organizational strategic objectives and your asset management plans even though you have a policy in place. A SAMP is required to document how organizational objectives are converted into physical asset management objectives and deliver the physical asset management policy."

"As mentioned in Neel's paper," Darren continues, "many organizations have been omitting this component and jumped right to developing and implementing their physical asset management plans. This is a formula for failure, and we have seen how organizations are struggling to operationalize their plans."

"This means that a plan without a strategy cannot be implemented successfully and deliver the desired results," Jerry interjects while staring at Darren. "If I understand correctly, the SAMP provides the link and fills the void that exists in a physical asset management framework. In short, the SAMP helps achieve alignment in the organization. Is that correct?"

Darren agrees with what Jerry said: "Correct! " he responds and continues: "Alignment is such an important aspect in physical asset management. Even the best people and the best tools cannot guarantee an effective physical asset management system if there is poor or no alignment."

"Another very important aspect of a SAMP," Darren says, "is that it consists of two key elements, as you will find out in the draft paper: the physical asset management strategy and the physical asset inventory and conditions. These two elements form the strategic plan which will incorporate strategies to achieve the physical asset management objectives. As mentioned before, you will need to know at a high level what physical assets you have and what will be required to manage them over the long term. Then you will need a physical asset management strategy consisting of policies, processes, and procedures to get you where you want to be."

"Sorry, but I am a little unsure about the physical asset management strategy." Jerry says, perplexed. "Can you explain what it would consist of?"

"Sure," Darren quickly replies as he starts to get very excited with the topic. He swiftly glances at his watch and sees that it's already ten o'clock. "Or maybe we will not have enough time," he says in a disappointed voice.

Just as Darren finishes his sentence, he sees Debbie and Rita walk into the dining room. Debbie is holding little Mia in her arms. Mia has been sleeping tightly all this time they have been here. Both ladies were in the kitchen, but Mia must have woken up while they were chatting.

"Debbie, are you in a rush to leave, or can we stay for another half hour?" Darren asks, already knowing what the answer will be, as he sees her happily rocking Mia in her arms.

"That's fine," Debbie replies with a smile. "You can take your time, unless Rita and Jerry are tired and need to get some rest," she adds, looking at Rita.

"You can stay as long as you want," Rita replies.

As soon as Darren hears that, he turns his attention back to Jerry and continues their conversation: "Yes, so you are asking about physical asset management strategy. As mentioned before, this will define the business processes and procedures around how physical assets are going to be managed in the organization. It acts like the glue for all the different functional areas that run across the whole organization. The physical asset management strategy provides guidance and directions to functional areas to formalize decision-making processes, achieve consistency, and promote collaboration. It will consist of a comprehensive set of strategies needed at the core of your physical asset management system to manage all physical assets owned, operated, and maintained by the organization in a structured, consistent manner."

Darren pauses to eat a spoonful of the bread pudding and then continues: "The objective of these strategies is first to guide implementation of the physical asset management framework so that all physical assets are treated as essential components of an interrelated system, rather than as isolated parts. In other words, ensuring there is alignment. Second, the structured consistent manner will provide a common approach and collaborative processes for the different functional and service areas to collect relevant data to evaluate, enhance, and maintain physical assets. Third, establishing an integrated set of processes and procedures will help minimize the lifecycle costs of owning, operating, and maintaining physical assets at an acceptable level of risk while continually delivering the required levels of service. And last, the physical asset management strategy will help the organization in making informed decisions related to physical assets. Optimized decision making is the core of an effective physical asset management system: decision making using the appropriate tools and techniques that help do the right thing, at the right time, in the right way, at the right cost for today and tomorrow and to achieve the physical asset management objectives."

Darren turns to a blank page on the note pad that Jerry brought from the den and starts scribbling on it. After a few energetic drawing moves, he turns the pad to Jerry and adds: "These are the key strategies that I would recommend you work on with your functional and services areas to develop synergies for some structured and consistent business processes. The set of asset management strategies is shown in the diagram" (Figure 4.1).

FIGURE 4.1 Asset management strategies.

Jerry contemplates the sketch for a few minutes and looks up. "This looks interesting but also seems difficult to develop, let alone implement across the organization," he replies in a disappointed voice.

"Yes indeed," Darren answers right away to try and appease Jerry's worried tone. "The physical asset management strategies development is definitely a huge undertaking and will take quite a few tough years to implement. However, these strategies are an integral part of the organization's physical asset management system, and, once in place, they will reap lots of great benefits."

"Remember," Darren says, an effective physical asset management system is built on sustainable and continuous improvement principles which will allow you to have comparable information derived from reliable data to make informed and optimum decisions. Another key benefit is achieving integration and alignment of corporate priorities while reducing redundancies and inefficiencies. Your organization will be able to leverage streamlined decision making to maximize corporate investment, while making the necessary positive impact on its triple bottom line—economic, environmental, and social" (Figure 4.2).

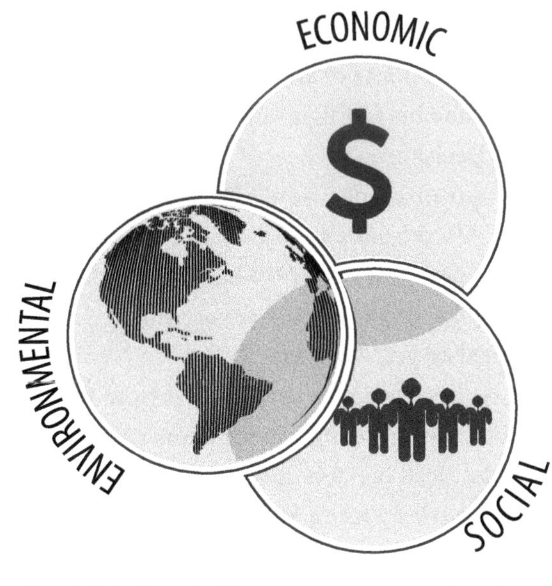

FIGURE 4.2 Sustainable asset management investments.

Jerry gives Darren a reassured look. With a poised voice he asks: "OK. So where should we start?"

"What was one of your selection criteria to buy your house?" Darren asks in response to Jerry's question.

Jerry looks at Darren as if he has seen a ghost. He must be thinking, "What has happened to Darren?!" for him to suddenly switch the conversation to his house.

"You heard it right," Darren continues while chuckling. "What were some or selection criteria for you to choose this house?" he asks again.

"Well, price of course," Jerry replies in a resolute voice. After a few moments of pondering, he adds, "And house layout for sure," while glancing at Rita.

He looks at Darren as if to ask "Is that good enough?"

Darren signals him to keep going.

After a few seconds Jerry exclaims, "Oh, and location!"

"Right on," Darren declares. "Location, location, location—that's one of the key deciding factors in real estate as 'data' is in physical asset management."

Jerry thinks for a moment that Darren's little charade is over, but he is wrong.

Darren continues, "What's another key factor to consider in physical asset management and that was also important in purchasing your house?"

Jerry looks dejected and does not know what to say. Suddenly his eyes stop on the sketch he is holding, and his face lights up with a big smile. "I think I know!" he answers with delight. "Decision!"

"Correct again—optimized decision!" Darren quickly responds. "You're a very smart chap, Jerry," Darren adds as he glances at his watch.

"I think it's time for us to leave now. We had a very good time, and that was a great dinner. We would like to thank you for everything."

"But we haven't finished yet," Jerry says.

"I know," Darren replies. He pulls a set of papers from his bag and hands them to Jerry. "Here. One of them is the draft paper I was referring to from one of my undergraduate students, Neel. It elaborates on physical asset management strategy. The other one is an article I wrote a few years ago while implementing an asset management program in my last company. I am sure you will find them very interesting and will help you continue with what you have started at Mouroum Inc."

Jerry takes both documents and gives Darren a firm handshake.

"Thank you so much, sir! I really appreciate your guidance and support."

"You're welcome, Jerry."

As soon as Debbie and Darren leave Jerry's house, it starts to drizzle. During the drive home, Debbie talks about Mia and how cute she is. She also reminds Darren of a few things Rémi and Lucas did when they were small, and they both laugh about those fond memories on the way home.

It's two o'clock on Monday afternoon, an unusually quiet day, as Jerry's boss, Robert Veristrik, the COO, is off on vacation. He will be away for the whole week. Jerry completed some key tasks since this morning, and he is now sitting in his office reflecting on what he can do next. Suddenly, he remembers the documents Darren gave him a week ago. He didn't have a chance to read through them, as he had a really busy week compiling a few reports for Robert, who needed them completed before leaving for vacation. Jerry has been carrying the documents with him in his bag to the office and back home all week but could never find time to review them.

Now is the perfect time. Jerry goes in the kitchen and makes himself a big mug of black coffee. He closes his office door so that he doesn't get disturbed and takes the documents out from his bag. First, he decides to read Neel's paper, as he's still unsure about the different strategies that form the physical asset management strategy in an organization's physical asset management system.

Jerry takes a long sip of coffee, settles comfortably in his chair, and starts reading:

> Your organization's physical asset management strategy consists of a few key strategies that need to dovetail nicely with each other to ensure alignment and provide that structured and consistent approach across the whole organization to meet the strategic physical asset management objectives. Below are some of the key strategies.
>
> ## Level of Service Strategy
>
> Whether you are in the business of producing goods or delivering services, your operations are constantly evolving in order to remain competitive and to be able to meet new expectations. This involves agreeing on and then

monitoring a set of defined technical and customer level of service measures. Organizations need to develop a framework and strategy to identify levels of service to customers both as expected by the customer and as delivered by the organization. Then it is a matter of determining what costs are required to maintain this level of service, which will provide a solid rationale for assessing the benefits/costs associated with enhancing specific product attributes or areas of service. Having a level of service strategy has proved to be a robust approach to physical asset management. It enables both the justification and prioritization of capital expenditure for the renewal, refurbishment, and replacement of physical assets that are associated with the delivery of required levels of service.

Fundamentally, achieving a set level of performance in the form of goods produced or services delivered is the core of an organization's existence. Physical assets exist for the purpose of supporting organizations to deliver those goods and services at the required level of performance to its customers, both internal and external. Level of service indicators in a nutshell reflect the degree of compliance with regulatory requirements, specification standards, and customer preferences or expectations. Physical asset performance indicators are drawn from the specific subsets of levels of service indicators and can be used to monitor the effectiveness of the physical asset maintenance, renewal, and replacement routines from year to year. The level of service indicators also help to support the development of the organization's strategic asset management plans and provide clear objectives and goals.

One key challenge of physical asset management is to optimize the balance between the required levels of service, cost, and risk. How to provide the required levels of service at optimum cost while minimizing risks? This entails not only a better understanding of customer expectations, but also a better understanding of the costs involved to deliver the levels of service and the risks and consequences of not delivering them. Typically, levels of service in an organization are developed at three different levels, and they are all linked together: customer, corporate, and technical (or asset). Levels of service identified at the customer level set the stage for organizations to develop and agree on the corporate service levels to be delivered, which are then translated into technical/asset levels of service. Ultimately, the level of service strategy allows organizations to evaluate the costs associated with delivering the set levels of service to assess affordability and competitiveness of the market/customer. This then helps determine the required resources as well as the funding strategies to deliver to the agreed levels of service.

In order to develop the levels of service for an organization's products or services, a structured framework must be followed (Figure 4.3). The elements of the framework are described below:

- Customer Level of Service (LOS) describes the products/services and level of performance that the customer expects (e.g., maximum xx minutes wait time at the departure gate).
- Corporate Level of Service describes the corporate objectives (e.g., providing excellent customer service at the departure gate).
- Technical or Asset Level of Service describes the technical requirements needed to achieve products/service objectives (e.g., available and reliable boarding bridge at the departure gate).

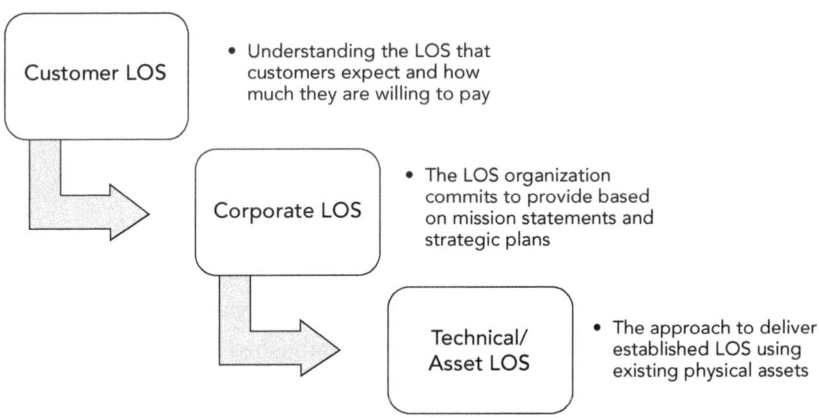

FIGURE 4.3 Level of service (LOS) framework.

The level of service framework helps to clearly document the customers' requirements, which is of utmost importance in order to understand and quantify the perception and value of the expected levels of performance by customers. At the center of providing those levels of performance are physical assets; hence, management of these physical assets needs to be considered against the customers' needs and expectations—how much can be achieved, at what costs, and how affordable will it be.

The level of service strategy is an important piece of an organization's strategic asset management plan. It allows decision makers to have established performance measures to base their decisions on and to understand the impact of those decisions on the levels of service being delivered. This can largely help organizations to formulate their operating and capital budgets as well as clearly communicate service impacts.

At minimum, organizations are required to:

- Diligently document levels of service for all three elements of the framework in a sequential manner in order to provide alignment between strategic organizational objectives and physical asset-focused performance.
- Develop, measure, set targets, monitor, and communicate performance measures for the documented levels of service.
- Regularly review and update the levels of service based on changes in customers' needs and expectations or in the organization's strategic direction.
- Operationalize the strategy by ensuring that the impact of the level of service is considered in all capital investment and operational decisions.

Risk Management Strategy

To assess the impact of levels of service in their decision-making process, organizations need to include a risk evaluation component. This is important to identify risks from all sources—not just physical asset-related, but also all risks related to failure to deliver established levels of service. As mentioned earlier, physical asset management involves understanding and balancing levels of service, cost, and risk. A risk management strategy is another important piece of an organization's strategic asset management plan to achieve those objectives. A risk management strategy aims to identify, assess, and manage risks while maintaining a strategic risk register. The register contain all major risks along with details of planned mitigating measures and the level of control that those measures provide.

The goals of a risk management framework should ensure that:

- All the different types of risks in an organization are understood and identified.
- The risks are individually assessed to determine their level of impact.
- The highest risks are prioritized so that they can be addressed in the short to medium term.
- Risk mitigation strategies and treatments are identified and applied.
- Organizational objectives are consistently met while managing risks.

As customer and corporate levels of service are translated to technical/asset levels of service, organizational and operational risks also translate to physical asset risks. Risk management should be viewed as an integral part of managing physical assets over their whole lifecycle phases to better understand the assets' inherent risk profiles. This helps in developing and implementing appropriate strategies so that risks can be managed to acceptable levels. For physical assets, risk mitigation can be achieved through the rehabilitation, renewal, and replacement of the assets or by monitoring the risk for actions to minimize the consequences. Risks that are considered unacceptable or have significant consequences are associated with their respective physical assets for further action. The first thing that organizations have to undertake is to develop a risk management strategy to ensure that:

- All critical physical assets within each of the service areas that are involved in the delivery of agreed-upon levels of service are assessed and documented.
- All significant physical asset risks are identified and categorized in a risk register based on established criteria, such as health and safety, environmental, and operational criteria.
- All identified physical asset risks are assessed in terms of both likelihood of occurrence and consequence.
- Physical asset risks are prioritized, and strategies for mitigating them are developed.
- Implementation of the physical asset risk management process is monitored and reported.
- Risks are considered in the decision-making process to determine physical asset management interventions priorities.

A risk assessment framework is based on the following simple and very common formula:

$$\text{Risk} = \text{Probability of Failure} \times \text{Consequence of Failure}$$

The probability or likelihood of physical asset failure provides an answer to the question "How frequently will the event occur?" This can be estimated on a scale from 1 to 5, with 1 being "very low" and 5 being "very high." The consequence of physical asset failure determines the severity of the failure on a scale from 1 to 5, with 1 being "low severity" and 5 "high severity." The product of the two scales allows for the quantification of risk.

Examples of probability and consequences of physical asset failure scoring categories are shown in Tables 4.1 and 4.2, respectively.

TABLE 4.1 Example of Probability of Asset Failure Scoring

Probability Score Category	Description of Occurrence
1	Very low—Hasn't happened before anywhere
2	Low—Has happened somewhere before
3	Medium—Can happen here
4	High—Has happened here before
5	Very high—Happens here often

TABLE 4.2 Example of Consequence of Asset Failure Scoring

Consequence Score Category	Description of Consequence
1	No impact on service delivery and operational costs
2	Low impact on service delivery; basic functional performance is available and acceptable
3	Service delivery is affected but with no major implications, as an alternative solution is readily available
4	Important to service delivery operations with significant impact; need quick action and special attention to address situation
5	Vital to service delivery operations and high profile with critical results; major impact and rigorous intervention required

Once a risk assessment is completed, there are a few options to be considered with regard to managing the risks:

- Transferring the risk can be achieved through the use of various forms of insurance, or through payment to third parties who are prepared to take the risk on behalf of the organization.
- Tolerating the risk is where no action is taken to mitigate or reduce a risk; either the organization is ready to accept the risk as is, or it is not cost-effective to take any action.
- Treating the risk is controlling it through actions that reduce the likelihood of the risk occurring or that minimize its impact prior to its occurrence.
- Terminating the risk is simply altering an inherent risk in a physical asset or process in order to remove the risk.

For each of these risks, a number of solutions are developed that either reduce the risk significantly or reduce it to a more acceptable level. A risk rating framework can help to map, prioritize, and navigate through the risks as they are monitored and managed (see Table 4.3).

TABLE 4.3 Risk Rating Framework

		Consequence				
	Definitions	**1 No Impact**	**2 Low Impact**	**3 Moderate Impact**	**4 High Impact**	**5 Significant Impact**
Likelihood of Failure	5 Very High	5 Moderate	10 Significant	15 Significant	20 Severe	25 Severe
	4 High	4 Low	8 Moderate	12 Significant	16 Significant	20 Severe
	3 Medium	3 Low	6 Moderate	9 Moderate	12 Significant	15 Significant
	2 Low	2 Low	4 Low	6 Moderate	8 Moderate	10 Significant
	1 Very Low	1 Low	2 Low	3 Low	4 Moderate	5 Moderate

As part of the physical asset risk management strategy, periodic reviews, updates, and monitoring of the risk register are essential. High-level reviews help to reasonably predict any impacts on corporate levels of service that could be expected to occur in the short term and long term. It is very important to identify potential risks, their likelihood of occurrence, and the potential consequences. A risk management strategy provides an understanding of the risk profile of the organization's physical assets base and helps to identify those physical assets that are most in need of attention. Focusing on the physical condition or the remaining useful life of a physical asset is not enough to determine its risk profile. To effectively operationalize physical asset risk assessments, the operating context of a physical asset also needs to be considered, as this can impact the rehabilitation, renewal, and replacement intervals. For example, exposure to extreme weather conditions due to climate change can accelerate the deterioration rate or create new failure modes on existing physical assets. Furthermore, the risk assessment needs to take a consistent risk-based approach to assess the relative priorities of individual physical assets before making any investment decisions. The approach will take into account potential issues and consequences and will endeavor to mitigate impacts to physical assets and levels of service in the short term and in the future, while considering the affordability factor as well.

Linking the outcomes of a risk assessment review with levels of service and costs while integrating climate change impacts will build a solid decision-making platform for the organization's budgeting process. This can support the timely allocation of funds to manage physical assets over their whole lifecycle.

Lifecycle Management Strategy

A comprehensive approach to physical asset management involves activities and practices for maximizing the performance of physical assets while minimizing risks and optimizing costs throughout their whole lifecycle. These lifecycle management activities allow organizations to maximize the realization of value from their physical assets from the planning, design, and acquisition phases through to the operations/maintenance phases before disposal (see Figure 4.4). Executing lifecycle management activities effectively enables organizations to make better decisions and to better plans for when and how much to invest on what for the short term as well as the long term.

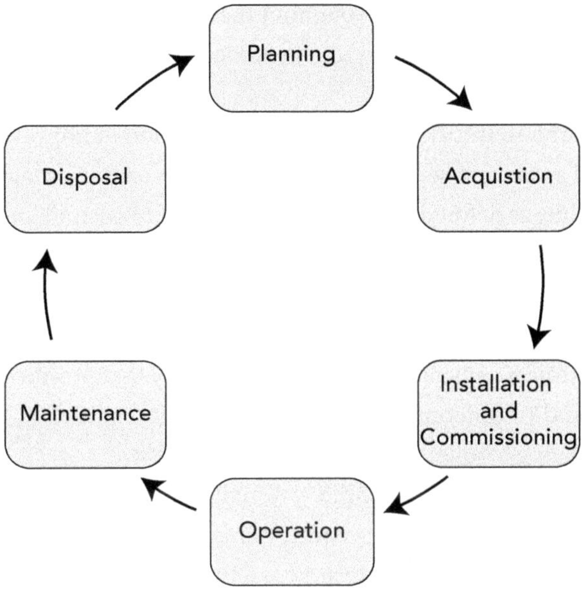

FIGURE 4.4 Physical asset lifecycle phases.

An organization's physical asset lifecycle management framework includes key methodologies/strategies that take place over the different lifecycle phases of physical assets. Some of them are listed below:

- SYSTEMS ENGINEERING—This interdisciplinary field of engineering and engineering management focuses on how to design and manage complex physical asset systems over their entire lifecycles. Systems engineering utilizes systems thinking principles with a holistic approach to deal with issues such as engineering requirements, reliability, maintainability, performance standards, testing and evaluation, and many other disciplines. Coordination with different cross-functional teams is necessary for successful physical asset system development, design, construction, operation, maintenance, and decommissioning.
- CONFIGURATION MANAGEMENT—A systems engineering process for establishing and maintaining consistency of a physical asset's performance, functional, and physical attributes with its requirements, design, and operational information throughout the asset's life. The process is

used to manage changes throughout the whole lifecycle of complex systems, such as in the military, civil engineering, and other industrial engineering segments.

- VALUE ENGINEERING—A systematic method to improve the value of physical assets and services thorough analysis and examination of function. Value, as defined, is the ratio of function to cost. Value can therefore be manipulated by either improving the function or reducing the cost.
- PROJECT MANAGEMENT—The application of knowledge, skills, tools, and techniques to plan, organize, and manage the effort to deliver a successful project. A project is a one-time activity that produces a specific output and/or outcome (e.g., building a bridge, expanding a facility, or rehabilitating a boiler system).
- OPERATIONAL STRATEGIES—Activities designed to ensure sufficient utilization of physical assets and systems. These are the regular tasks that are undertaken to ensure physical assets and systems achieve their service potential. Operations strategies include activities such as inspections, system monitoring, and emergency response planning.
- MAINTENANCE STRATEGIES—Approaches for determining and implementing the optimal mix of proactive, detective, corrective, and reactive maintenance tactics. Maintenance strategies are designed to enable existing physical assets to operate to their full service and performance potential over their entire useful life.

A key element of lifecycle management strategy is achieving the most effective lifecycle cost of physical assets over their entire life. Whole-life costing is a process to determine the sum of all the costs associated with both capital and ongoing operating and maintenance interventions, including acquisition, installation, operation, maintenance, refurbishment, and disposal costs. Typically, the acquisition costs of an asset (i.e., planning, design, and construction) represent a relatively small proportion for the total cost of ownership, with the total operating and maintenance costs being a more significant amount over the useful life of the asset, as shown in Figure 4.5.

However, in many instances, some of the operating and maintenance costs incurred are the result of substandard designs or poor decisions during the physical asset system acquisition phase. Hence, it is important not to be shortsighted but to make decisions with the long-term goals in mind. The overall operations and maintenance strategy can only maintain the current levels of service based on the inherent design capacity and mitigate risk as much as possible while minimizing cost. One of the biggest challenges in a physical asset lifecycle management strategy is to come up with the most cost-effective blend of planned and unplanned maintenance activities once physical assets are put into operation.

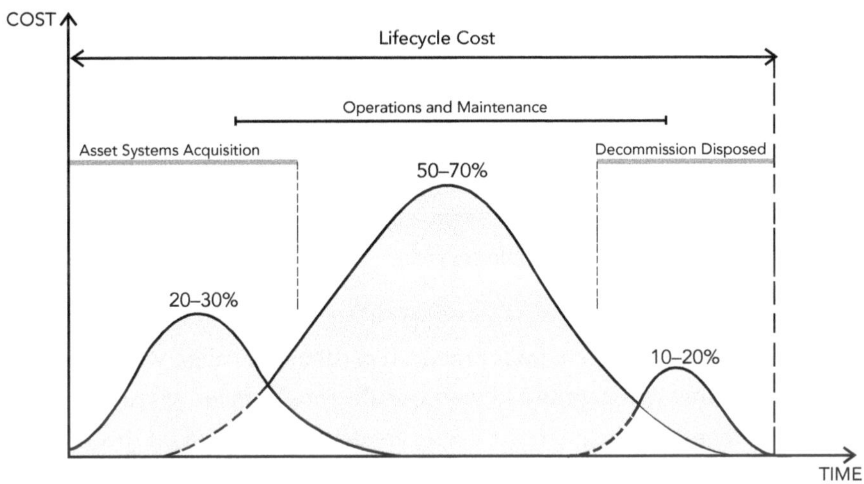

FIGURE 4.5 Distribution of physical asset lifecycle costs.

Demand Management Strategy

Predicting and finding optimum means to manage current and future demands for services and performance levels are essential to an effective asset management strategy. Many factors and trends such as population growth and social and environmental changes can influence demand. Demands typically change over time and are usually related to many different variables. Understanding these variables is important to develop predictive analytics and forecast future demands. Demand analysis does not always result in increasing products or services. For example, a population increase can mean more frequent or reliable transportation services, while a change in consumption trends may show a decrease in usage of a certain product.

Therefore, the impact of these trends and changes must be analyzed to assess the impact on future demands. This will dictate the demand management strategies recommended to proactively manage those demands as required without compromising customer satisfaction or service performance.

Demand management strategies provide alternatives to the acquisition of new physical assets in order to meet demands and explore other means to influence demands. For example, the effect of climate change with hotter summer temperatures does not always mean upsizing the air conditioning system. Other solutions such as planting more trees to create shade, limiting the internal heat loads, or adopting passive cooling strategies can help. The drivers for implementing demand management strategies typically include:

- Optimize utilization/performance of existing physical assets.
- Reduce or defer the need for new physical assets.
- Meet operational cost-efficiency objectives.
- Deliver a more sustainable service.
- Respond to changing needs.

Demand management is practiced constantly to align with the capital investment process and to maintain the total demand at reasonable levels. Demand management strategies are developed based on specific demand components to target specific applications, as shown in Table 4.4.

TABLE 4.4 Examples of Demand Management Application

Demand Component	Demand Management Strategy	Examples of Application
Regulation	Rainwater collection	Require all new condominiums to have a holding tank of a certain capacity
Operations	Optimize asset usage	Improve asset reliability and availability
Incentives	Energy efficiency	Provide park-and-ride facilities to reduce GHG emissions
Education	Operational safety improvement	Train operators on existing assets to reduce safety incidents
Demand substitution	Frequent garbage pick-up	Provide larger garbage bins

Optimized Decision-Making Strategy

A key component of an organization's asset management strategy is an optimized decision-making strategy. This includes techniques based on risk evaluation, multi-criteria analysis, and approaches to optimize investment across the organization. An optimized decision-making strategy provides a robust, repeatable, and defendable investment planning process for the prioritization of projects for different physical asset classes.

Informed physical asset management decisions are required for:

- Operation and maintenance interventions
- Project solutions selection
- Project ranking within a department
- Project prioritization across departments

Optimized decision making, either within or across departments, must be based on a range of approaches which utilize the available physical asset data, such as condition assessment information, risk assessment information, expected levels of service, physical asset performance, and lifecycle costs. All this is supplemented with expert knowledge from staff across the organization and supported by professional recommendations from external parties. The optimized decision-making process includes more advanced approaches, such as needs assessment and assessment of a range of alternative solutions to select the appropriate one. The decision process involves assessing a broad range of capital solutions, such as renewal, rehabilitation, and replacement options in addition to operational solutions, such as enhanced maintenance programs. Ultimately the decision made will be the most appropriate to the organization's needs while maintaining a balance between levels of service, risk, and cost.

The decision-making process will involve a combination of risk-based analysis (RBA), cost-benefit analysis (CBA), and multi-criteria analysis (MCA).

Risk-Based Analysis

Risk is one of the key factors that needs to be considered in all aspects of optimized decision making. A risk management strategy, as explained earlier, quantifies the risk, identifies mitigation measures, and then sets out to reduce the risks in the most cost-effective manner. A key trend applied by many organizations is the integration of climate change impacts in the risk assessment process. RBA takes the outcomes from the risk management strategy and focuses on risk reduction while maintaining minimum cost.

Cost-Benefit Analysis

Cost is another key factor in the optimized decision-making process. CBA involves identifying the financial impacts of various alternatives within a business case. This includes both benefits and costs over the entire analysis period. The goal is to assess which alternative presents the greatest value of benefits compared with costs. This involves better defining the needs backed with robust data, thorough review of all potential solutions (including the "do nothing" option), and precise cost estimates, and also ensuring the solutions can be implemented and will achieve the required benefits.

These days there are growing concerns about how cost estimates of project alternatives are determined, evaluated, and considered in the decision process. Traditionally, the focus has always been on the initial capital cost estimates. However, there is growing pressure to have more accurate lifecycle cost estimates by including operating and maintenance costs, which are relatively higher over a physical asset's service life, as mentioned earlier. Lifecycle cost estimates are critical input to the evaluation of different project alternatives and in the decision-making process. They help to balance capital with operating costs in order to achieve an optimum whole lifecycle cost over a physical asset's lifetime. For example, when evaluating different asset intervention options, the total cost of projects including capital, operational, and maintenance expenditures discounted to net present values must be considered before making any decision. This will prevent situations such as where a cheaper capital solution with significantly higher operational costs is chosen over a slightly higher capital cost solution with significantly lower operational costs.

Multi-Criteria Analysis

The MCA approach typically utilizes a set of service/performance attributes which reflect the organizational strategic goals and objectives. This provides alignment across the organization to help determine which combination of projects represents the best-value capital program based on the current agreed-upon levels of performance.

The MCA approach takes into account the level of service strategy, the different service attributes, and their respective performance measures. It also considers the *triple bottom line* (social, environmental, and economic) factors.

For each project need, a score is attributed to each service/performance attribute that best reflects the value provided by the project under consideration. The collective scores and weightings for the project are then combined to determine its alignment and total benefit value. This allows for ranking and prioritizing of projects and helps decision makers in determining the best-value capital investment program over the forthcoming budgetary periods.

Knock, knock, knock.

The sound startles Jerry in his chair. He was so focused on reading Neel's paper that he completely lost sense of everything. He realizes somebody is knocking on his office door.

"Yes, come in, please," Jerry responds in a shaken voice.

It's his assistant, Rebecca. She informs him that she is leaving and that everything is all taken care of.

"See you tomorrow," Rebecca says as she smiles and leaves, closing the door behind her.

Jerry looks at his watch and notices that it's already four o'clock. He has been reading for almost two hours. Neel's paper was so interesting that he was completely immersed in it. He remembers the last discussion with Darren regarding the best place to start the physical asset management program, which is to tackle the decision-making process. Neel's paper does provide the answer in the form of the physical asset management strategies that need to be in place to provide the required alignment in organizations, an essential component before even thinking of executing any plans.

Jerry flips through the last few pages of the paper and puts it back on his desk. He gives a big sigh, puts his two feet on his desk, and stares at the ceiling.

"What next?" Jerry thinks to himself.

He looks at his watch again and pauses for a few seconds.

Then he grabs his cell phone and calls Rita.

Nobody picks up, and it goes to voicemail.

"Babe, I will be a couple of hours late tonight. I have a few things to finish. I should be home before dinnertime. Love you."

Jerry hangs up and ponders how Rita will take it that he will be home late once again.

He has been doing this for quite awhile now, as the new job and higher expectations have taken a toll on him. Rita understands that he has to work hard, but she is more worried about Jerry's health and insists that he has to have a balanced lifestyle.

But all is well, Jerry thinks to himself. This is only temporary, and, in a few months, things will start falling into place. On top of that, he knows that Rita trusts him and that he has her unwavering support.

Jerry goes to the kitchen again and gets himself another cup of coffee. He uses the opportunity to stretch, as he has been sitting for awhile. He looks out through the window of his office building overlooking the parking lot. There are still a few cars around, indicating some others are also working late. However, most have left already, maybe because Robert is not around.

CHAPTER 5

PHYSICAL ASSET PLANNING

Once again Jerry makes himself comfortable in his office chair and grabs Darren's article from the pile of papers. The article is a couple of pages long with diagrams. Jerry is excited to dive into Darren's brain (again) via this article and is eager to learn how Darren tackled physical asset management at his workplace. He notices the article's title is "Physical Asset Planning," and it talks about physical asset inventory and condition assessment. Jerry takes a deep breath as if he is preparing to complete an underwater free dive and begins to read:

> To develop a physical asset management strategy and plans, you need to know more about your physical assets and the condition they are in. That is to say, you need to take a snapshot of your current physical asset management stock—this includes the age, condition, and remaining life of the stock—to better understand what it will take to maintain it all in a "state of good repair." I call this process "developing the organization's physical asset plan," as it is more than an inventory with condition assessment data. In other words, the physical asset plan is sort of lining up all your ducks before you start making decisions about what to do with them, why, and when. In general, the physical asset plan is intended to provide information to answer the six following key questions:
>
> 1. What do you have?
> 2. What is it worth?
> 3. What condition is it in?
> 4. What do you need to do to it?
> 5. When do you need to do it?
> 6. How much money do you need?

What Do You Have?

An inventory database of physical assets covered by the plan includes a basic record of physical asset information, typically held in some sort of database or software system. It also contains physical asset attribute data, such as asset class/group, physical description, location, quantity, type, construction or acquisition date, and cost.

What Is It Worth?

Determining the valuation of a physical asset is important in order to carry out effective planning. This valuation is different from the financial accounting valuation that is done on an annual basis in organizations to meet financial reporting requirements. Financial accounting asset valuation involves applying a depreciation method to estimate the amortization, which shows that many physical assets that are beyond their useful lives may have been fully depreciated. Hence their net book value will be zero. However, from a physical asset management perspective, the physical assets continue to operate and perform to provide the required levels of service.

For physical asset planning and future replacement decisions, the financial approach is not appropriate, as it is based upon historical cost, and not the current cost to replace the physical asset. From a physical asset management perspective, it is therefore more effective to evaluate the physical assets using replacement cost valuation. The replacement values provide a more accurate estimate of the future cost required to replace the physical asset at the end of its useful life.

The replacement cost valuation is based on using a combination of the following industry practices:

- Escalation of original acquisition cost to current year
- Industry standards or market unit cost indices
- Review of recent historical contracts and tender pricing documents
- Condition assessment replacement costs based upon third-party cost estimates
- Property insurance values

What Condition Is It In?

The condition of each physical asset is evaluated to represent the current "health" of the organization's physical assets. The condition of physical assets is assessed from three different angles: physical, demand, and functional.

Physical condition reflects the physical state of an asset and is mainly related to the age and remaining useful life of the asset. Physical condition may or may not affect the performance of physical assets to deliver the required level of service. Some typical physical asset performance standards are reliability, availability, capacity, and specific operational requirements. Knowing the performance and physical condition of assets is valuable information to determine the remaining useful life of a physical asset and, more importantly, the timing for required interventions to bring the level of service to the desired performance standard.

Demand condition assessments examine physical asset capacity over the short and long terms. Physical assets must be utilized effectively in order to provide the maximum return on investment and to deliver the required levels of service or performance standard. Wherever possible, the goal should be the optimum utilization of physical assets to meet current levels of performance as well as plan for future requirements. This can be achieved through physical asset rationalization based on demand forecasts to improve operational efficiency.

Functional condition assessments deal with the suitability or "fitness for purpose" of the physical asset. Suitability needs to recognize service needs for current and future purposes. When assessing the suitability of a physical asset, opportunities to vary the level of service, and hence the cost of service, are considered.

Condition is rated using a five-point scale, which allows for comparative benchmarking between different physical asset classes and is sufficiently detailed for high-level decision making.

The rating scales for the three types of condition assessment are shown in Table 5.1.

TABLE 5.1 Types of Condition Assessment and Rating Scale

Condition Category	Description	Definition
Physical	Physical deterioration of the asset	**Very good:** Well maintained and fit for the future. New or recently rehabilitated.
		Good: Good working condition and adequate for now. Asset shows only minor deterioration.
		Fair: Some elements show signs of deterioration and require attention. Tolerable physical condition.
		Poor: At risk. Approaching end of service life and major deterioration evident.
		Very poor: Unfit or unsafe for sustained service. Near or beyond expected service life.
Demand	Asset's capacity to deal with long-term demand or usage	**Very good:** Fit for the future. Can comfortably deal with long-term demand, loading, or usage requirements.
		Good: Adequate for now. Can comfortably deal with medium-term demand, loading, or usage requirements.
		Fair: Requires attention. Current design and function meet requirements with some minor shortcomings.
		Poor: At risk. Operating at design capacity to meet short-term demands, loading, or usage requirements.
		Very poor: Unfit for sustained service. Operating at or below design capacity. Unable to meet short-term demands, loading, or usage requirements.
Functional	Alignment with the current functional needs	**Very good:** Adequate design and functions to fully meet current requirements.
		Good: Design and functions meet current requirements.
		Fair: Design and functions somewhat meet current requirements with minor shortcomings.
		Poor: Design and functions partially meet current requirements with significant shortcomings.
		Very poor: Design and functions barely meet current requirements.

What Do You Need to Do to It?

Over time physical assets deteriorate from condition rating 5 down to 1. The rate of deterioration is illustrated on a physical asset deterioration (or degradation) curve, as shown in Figure 5.1. One key point to note on the curve is the accelerated rate of deterioration toward the end of a physical asset's service life.

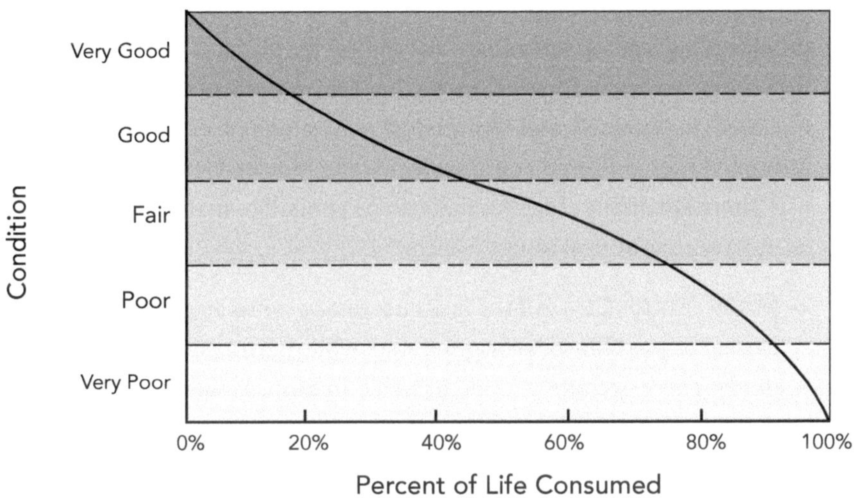

FIGURE 5.1 Physical asset deterioration curve.

The estimated percentage of the remaining useful life of a physical asset, based on the age of the physical asset, is considered a good starting point to estimate its overall physical condition.

Consider a truck. It might last ten years used on rural and urban roadways. However, if the truck is used at a mining or construction site, it likely would not last ten years. Operating context and usage of a physical asset impact the estimate of useful life. If usage is more or less consistent across an asset base, then the useful life estimate is a good tool to predict when that physical asset will need to be refurbished or replaced and can be used to estimate remaining value.

Infrastructure assets with longer useful lives undergo a continual process of repair, rehabilitation, and refurbishment in order to maintain their intended purpose. For example, roads, bridges, and facilities usually undergo continual maintenance and rehabilitation, and, hence, age may not be the most useful indicator to use for physical asset management planning. In many cases the physical asset useful life needs to be supported with other information such as condition assessments, operating context, history of failures and upgrades, lifecycle costs, demand, and growth forecast and expert judgment.

Estimated remaining useful lives, based purely on age, can sometimes provide a misleading forecast on the timing for physical assets' rehabilitation or replacement. Physical assets that are properly designed, constructed, operated, maintained, and refurbished may remain useful beyond their estimated useful life and continue providing required levels of service.

There are different tactics that can be applied to maintain or extend the useful lives of physical assets (Figure 5.2).

- MAINTENANCE—All the tasks necessary for retaining a physical asset as near as possible to its original state while performing to required standards. These exclude renewal or rehabilitation. Maintenance slows down the deterioration of physical assets to achieve their estimated useful life at optimum costs while managing the consequences of failures and delivering required levels of service. It also helps avoid premature rehabilitation or replacement.
- REHABILITATION—Tasks to rebuild or replace parts or components of a physical asset to restore it to as-new condition. In the process the remaining useful life of the physical asset is extended and the value of the physical asset increases.
- REPLACEMENT—This involves the complete replacement of a physical asset that has either reached the end of its useful life, is obsolete, or does not meet functional requirements in its current operating context. The new physical asset delivers a similar or agreed-upon alternative level of service.

All the abovementioned tactical interventions are very important and should be done at the right time and at the right intervals over the entire useful lives of physical assets.

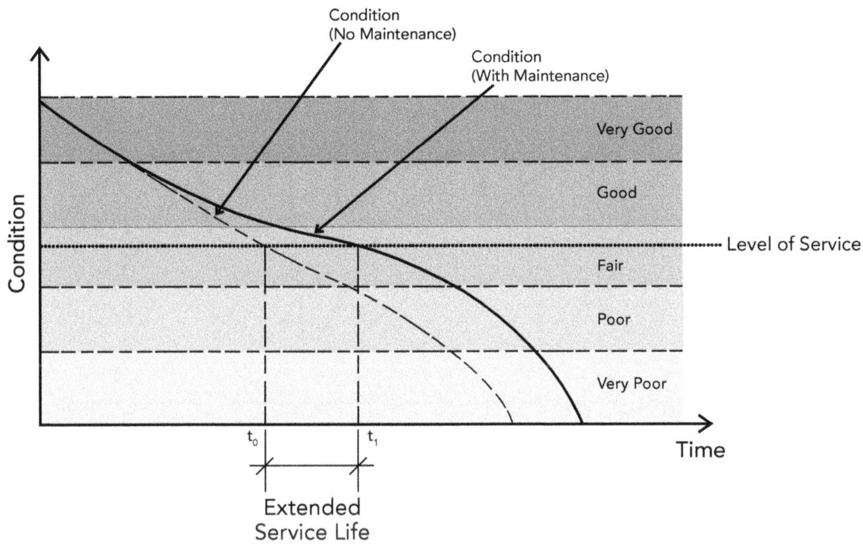

FIGURE 5.2 Asset maintenance to extend service life.

When Do You Need to Do It?

One of the essences of effective physical asset management is to ensure that the right maintenance, rehabilitation, or replacement is done at the right time and for the right reasons. The nature and frequency of these tactical interventions on physical assets depend on a few factors such as operating context, deterioration rate, criticality, technological obsolescence, redundancy, or climate change effects.

All are done to maintain a certain level of service and to minimize risks at optimum costs.

Deferring maintenance, rehabilitation, or replacement of physical assets can have huge consequences depending on the criticality of the physical asset and the operating context. Consequences can include not only decreased levels of service, but also health and safety risks, such as injuries and fatalities; environmental risks, such as impacts on the ecosystem, huge fines for noncompliance; or significant operational costs involving other secondary damages.

Yes, at the end it all boils down to costs to carry out those interventions (Figure 5.3).

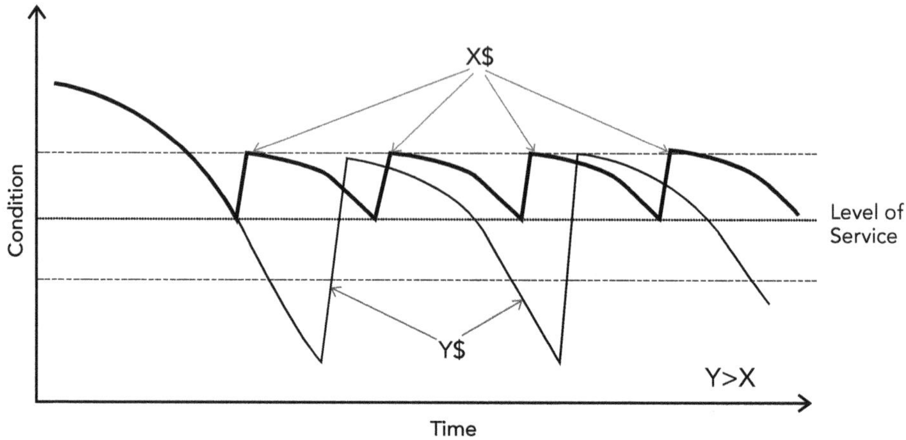

FIGURE 5.3 Deferring interventions consequence.

How Much Money Do You Need?

An organization is financially sustainable if it is able to maintain not only its financial capital asset but also its physical capital asset over the long term. To achieve that, an organization needs to have a good understanding of what will be required over the long term to maintain its physical assets in a state of good repair to deliver to the desired levels of service and performance standards. All the costs associated with the regular maintenance (operational costs), rehabilitation, and replacement (capital costs) required on the physical assets need to be mapped out over the years to forecast the funding required on an annual basis. These costs are illustrated in Figure 5.4.

Long-term financial forecasts for future operational and capital expenditures (maintenance, rehabilitation, and replacement) for all physical assets need to be determined as accurately as possible. In Figure 5.5, the forecast for capital expenditures is illustrated. The forecast highlights the current backlog (value of physical assets which are at or beyond the end of their useful lives) and the average annual capital investment requirements for the forthcoming years. It also shows the magnitude of the gap between the current and required capital investments and how the gap accumulates over time. From a physical asset planning perspective, this allows the organization to map out when physical assets need to be worked on and how much fund-

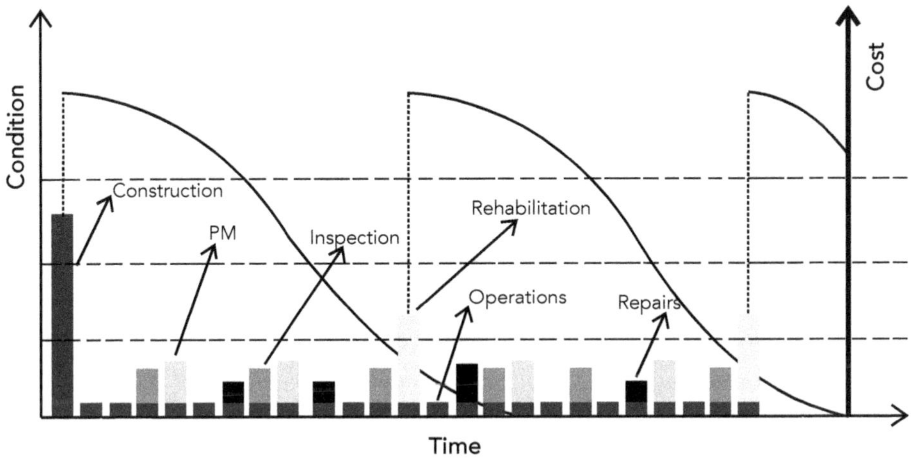

FIGURE 5.4 Physical asset intervention costs.

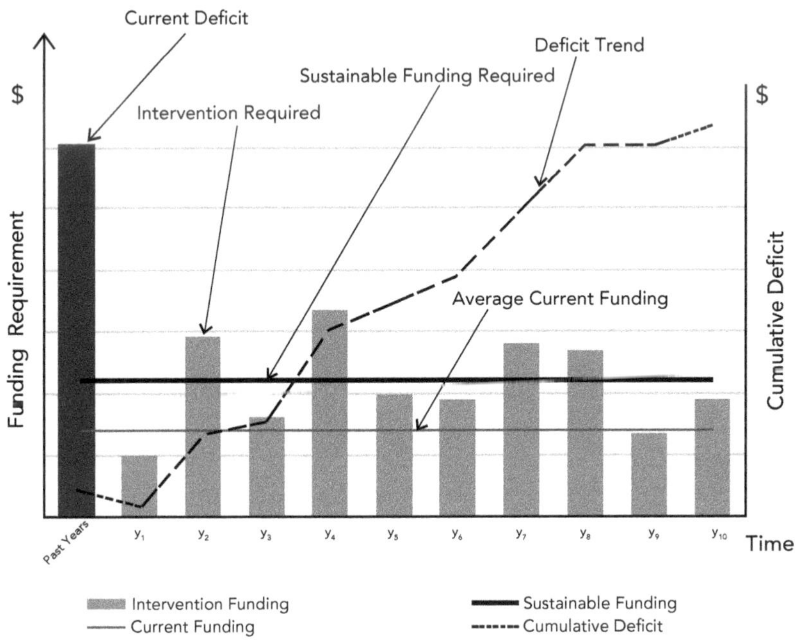

FIGURE 5.5 Funding forecast and funding gap.

ing will be required over time. This greatly helps those people in the finance department to carry out their own planning in their long-range financial plan to ensure that the necessary funds are available when required to manage the investment gap.

Jerry pauses for a couple of minutes to reflect on what he just read. He remembers what he read in Neel's paper about the process to follow in developing a strategic asset management plan. It mentioned that to develop a plan you need to apply the strategy to "something with a goal." The "something with a goal" is the physical asset stock and the value the organization wants to derive from it. This means that in advance the organization needs to know the physical assets it owns, their current conditions and risks, the estimated service life, and the estimated rehabilitation and replacement costs required to maintain them in good order.

Jerry's face lights up with a huge smile.

"All of this makes sense now," he utters to himself.

CHAPTER 6

THE TACTICAL ASSET MANAGEMENT PLAN

Once again, Jerry's legs go back on the table as he continues reading the article:

Once organizations have taken a snapshot of their current physical assets, they have a pretty good idea of their condition and a better understanding of the backlog they face as well as *what* needs to be done and *when* to their physical asset portfolios. The next step for organizations is to apply their physical asset management strategies to develop action plans to consistently meet the agreed-upon strategic asset management objectives.

So what are those actions that need to be planned to achieve the long-term goals of the physical asset management strategy? Well, organizations develop sets of tactical plans focused on short-term efforts based on resources available and practicality to achieve the strategic asset management objectives. Through their strategic asset management plan, organizations have set their asset management objectives with the deployment of a set of physical asset management strategies. They also know what situation they are in with their current stock of physical assets and what it would take to achieve those strategic objectives. Now is the time to start developing the tactical plans to execute the strategic asset management plan.

This is achieved through the development of physical asset management plans for each physical asset class. Each physical asset class will have its own set of plans consisting of operational, financial, capital, and maintenance plans. Let us take a look at each one of those plans.

Operational Plan

An operational plan consists of processes and activities that need to be performed to derive maximum value from physical assets to meet the organization's operational strategic objectives. It documents how physical assets will be operated on a day-to-day basis to produce the goods or deliver the services to the agreed-upon performance standards in a consistent and efficient manner. Operational strategic objectives translate the higher-level organization's strategic objectives and physical asset management objectives into operational performance targets and technical objectives. This provides the basis for the development of operational parameters and intervention levels, operational standard, operating practices/procedures, and operational decisions, both in normal operating circumstances and in emergency situations. An operational plan includes an emergency preparedness plan with incident responses, contingency plans, and business continuity plans in case of situations of emergency or crisis.

Financial Plan

Finance is a huge constraint when it comes to investment in physical assets. Long-term physical asset investment forecasts provide insight into prospective investment requirements which may fall outside of the 10-year planning horizon typically used in capital budgeting. Large quantities of physical asset acquisition during different points in time over the years will equally require heavy investment once those physical assets and their components start to reach the end of their useful lives. If those investment requirements are not addressed appropriately, levels of service could potentially decline, and operations and maintenance costs could increase. The multi-year forecast aims to cover the entire lifecycle of the different classes of physical assets, therefore allowing identification of trends for those future funding and investment requirements. The process to establish future average annual operational and capital costs needed in the long term was discussed earlier. This long-term forecast for physical assets investment has to be included in the organization's long-range financial plan. Long-range financial planning combines long-term financial forecasting with strategizing. Besides physical asset expenditures, organizations also have other assets to manage, and they must have a finance strategy in place to ensure the right amount of funding is available at the right time in the most effective manner. Long-term financial planning is a highly collaborative process that considers future scenarios and helps organizations navigate challenges to project revenues and expenditures over a long-term period, using assumptions about economic conditions, future spending scenarios, and other salient variables. Long-term financial planning is the process of aligning financial capacity with long-term physical asset management objectives to achieve long-term sustainability. It provides insight into future financial capacity so that strategies can be developed in light of the organization's strategic objectives and financial challenges.

Capital Plan

Capital investment planning is an important step in the development of an informed capital plan to meet the physical asset management objectives. As part of the optimized decision-making strategy in their strategic asset management plan, organizations have a range of decision-making techniques at their disposal. This is to help decide on and select optimal solutions for capital projects. The capital investment planning process applies the outcomes of these decision-making methodologies to develop a capital investment program. The capital investment planning process includes identification and scoping of the projects to address rehabilitation of existing physical assets and maintain the current levels of service, to renew or replace physical assets and address changes in levels of service or growth, and to acquire new physical assets as part of new capital investments.

The overall capital investment planning process is designed to support and align with the organization's strategic plan and objectives. This includes physical asset planning, acquisition, installation, and commissioning with the focus on linking capital investment to achieve the required levels of service, as shown in Figure 6.1. All departments making a capital investment request will follow the capital investment process with regard to identifying strategic goals as defined by levels of service, carrying out a need analysis to better understand the need, assessing a range of solutions (physical asset-related and nonphysical asset-related), prioritizing the agreed-upon solutions, and devel-

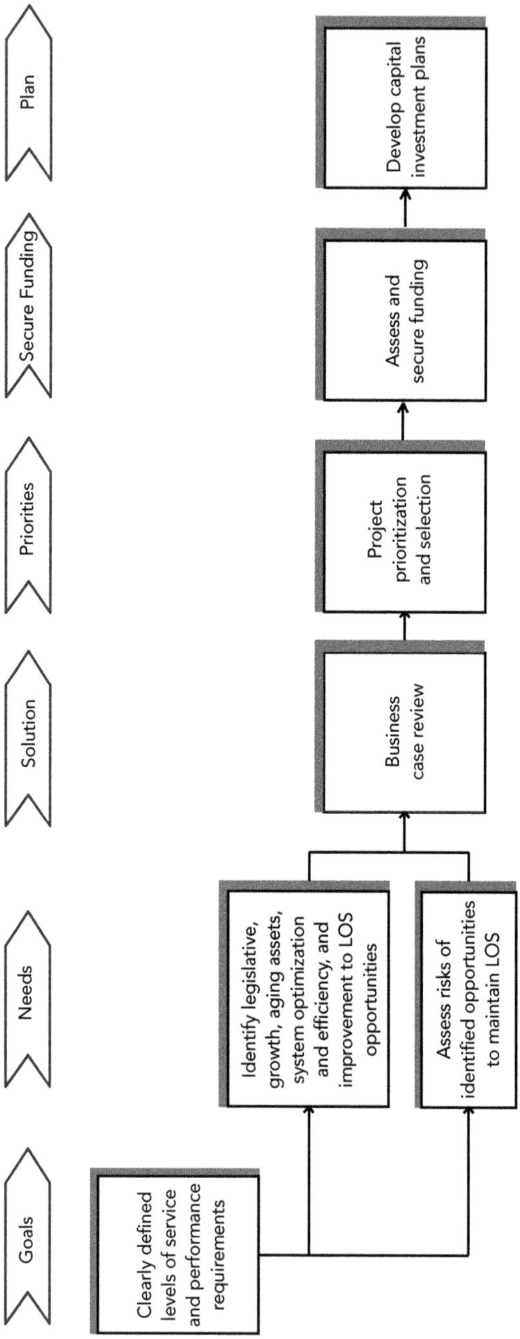

FIGURE 6.1 Capital investment planning process.

oping a capital investment plan. An organization's capital investment planning process can become a very tedious and complex task as multiple capital needs for physical assets from different departments are compared and prioritized in the face of the availability of limited funds. Hence, a consistent approach has to be adopted that will not only enable a more robust project prioritization process (see Figure 6.2) but also facilitate informed discussions on levels of service, risks, and long-term funding priorities across the different departments.

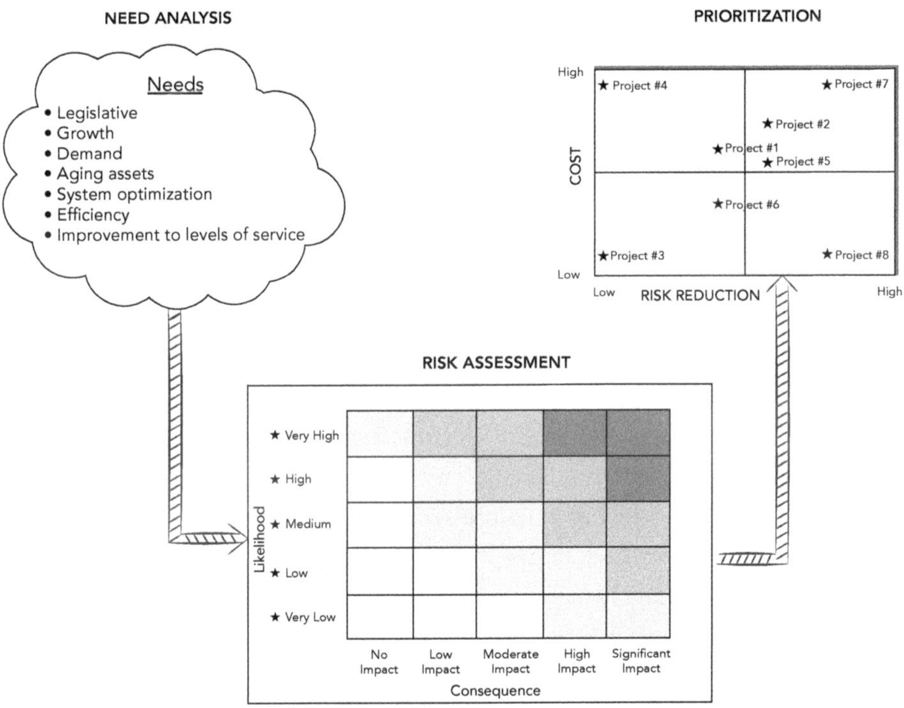

FIGURE 6.2 Project prioritization process.

A capital investment planning process helps an organization's senior management, shareholders, and decision makers in general to make better informed decisions with regard to capital investment. This would result in a capital plan that is well aligned with the organization's strategic plan and supports the physical asset management strategic objectives.

Maintenance Plan

Similar to the capital investment planning process, a maintenance planning process is an important step in an organization's physical asset management plan. As a reminder, a capital plan captures the capital investments needed to rehabilitate, renew, or replace a physical asset over its entire useful life. On the other hand, a maintenance plan captures the operational expenditures required to ensure that physical assets provide the required levels of service over their entire useful lives. Traditionally, organizations develop their maintenance plans using different approaches and methodologies. Similar to the capital plan development process, a maintenance plan must be developed with a focus on levels of service, performance standards, and risks all aligned with the physical asset management strategic objectives and organizational strategic plan.

A common methodology that provides such alignment when developing a maintenance plan is the reliability-centered maintenance (RCM) methodology and its seven basic questions.

1. What are the functions and associated performance standards of the asset in its present operating context?
2. In what way does it fail to fulfill its function?
3. What causes each functional failure?
4. What happens when each failure occurs?
5. In what way does each failure matter?
6. What can be done to predict or prevent each failure?
7. What should be done if a suitable proactive task cannot be found?

The seven RCM questions force those involved in the process to understand the context of the organization and operations, to define levels of service and desired performance standards, and to evaluate risks and consequences from different perspectives (i.e., health and safety, environmental, operational, and even nonoperational). Only then are action plans developed to manage those consequences. The beauty of the RCM process is that not only does it deliver a sound maintenance plan as outcome, but it also contributes to the operational and capital plan by identifying tasks that need to be undertaken in those areas based on impact to levels of service and performance standards.

Two important aspects worth mentioning when developing a physical asset maintenance plan are physical asset criticality assessement and root cause failure analysis.

Physical Asset Criticality Assessment

Organizations need to have a good understanding of all their physical assets and develop a means to analyze their significance to the organization. They might classify various physical assets in terms of the impact the physical assets would have when they fail. An organization-wide consultation process should be used to ensure the categorization process reflects the perceived criticality across all functions and departments. Different organizations might develop different criticality classification systems, as shown in Table 6.1.

TABLE 6.1 Physical Asset Criticality Classification

Risk Rating	Impact of Failure of Physical Asset
5	Major environmental, health, and safety disaster, public relations disaster, major financial loss, major damage to brand.
4	Moderate environmental, health, and safety impact, severe impact on stakeholder confidence, very high financial loss, damage to brand, international media coverage
3	High operational impact, high financial loss, stakeholder concerns, national media coverage
2	Moderate operational impact and financial loss, no impact on stakeholder values, local media coverage
1	Low financial loss, no impact on stakeholder values, no media coverage

Root Cause Failure Analysis

No maintenance plan is perfect at the outset. Operating context changes, and other internal or external factors may impact the performance of physical assets. A consistent approach in analyzing failures, identifying the root causes, and developing and implementing action plans can help continuously improve an organization's maintenance plan.

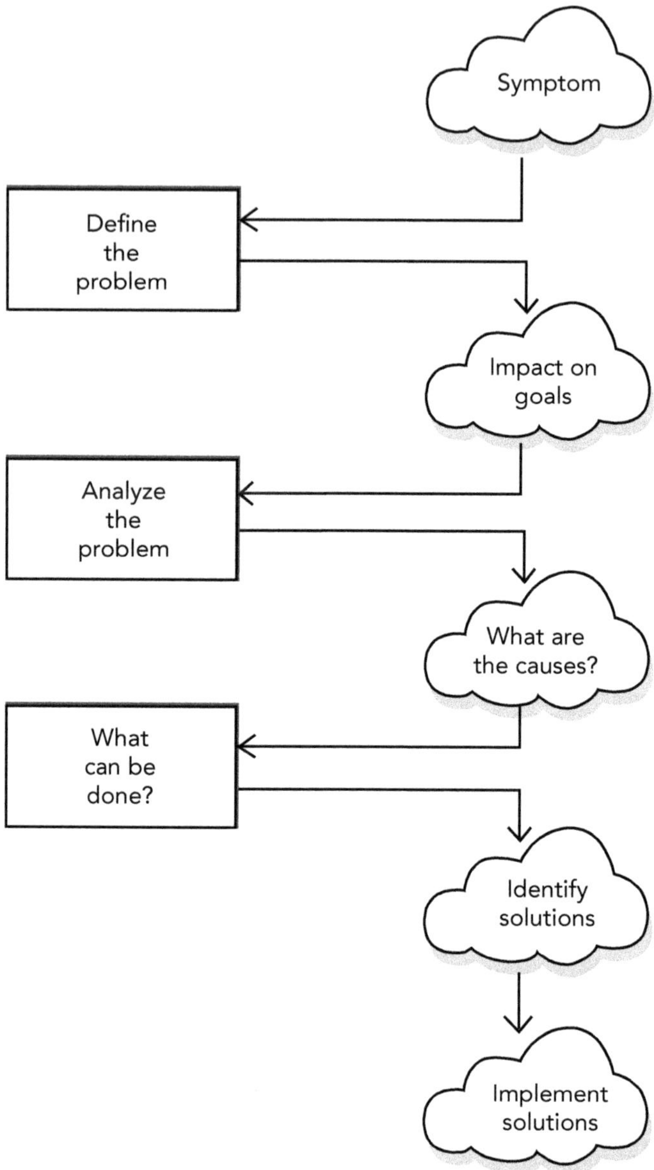

FIGURE 6.3 Root cause failure analysis process.

This process is especially important when an organization executes its capital plan. This means rehabilitating, upgrading, and replacing physical assets. These interventions may create new failure modes that have not been identified and taken care of in the existing maintenance plan.

Jerry looks at his watch and notices that it's already six o'clock and getting pretty dark outside. He puts the article down and thinks it's time to head home before Rita gets worried and starts calling. Nonetheless, he is happy with the few extra hours spent in the office reviewing Neel's paper and Darren's article. Both documents proved to be helpful in enlightening his understanding of different physical asset management approaches. Jerry knows he has obtained plenty of actionable information to improve his current physical asset management program. He takes a couple of minutes to reflect on what he read—the different strategies of a strategic asset management plan and the components of physical asset management planning.

As Jerry prepares to leave the office, he considers linking up with Darren in the coming days in order to get some direction on how to go about implementing what he read.

CHAPTER 7

ESSENTIAL ENABLERS

Darren recently finished delivering an exhaustive two-hour lecture on physical asset management. The audience consisted of graduate students eager to understand and learn more about the physical asset management discipline. The lecture was followed by an interactive discussion of about forty-five minutes with the students who were showing an engaged interest in physical asset management and wanted to better understand the requirements needed to enter the job market. There were plenty of questions focused on the job prospects in physical asset management, the required skill sets to secure a job in that field, and its main challenges.

It's ten minutes to noon, and Darren is reflecting on the questions asked before he takes his lunch break. Darren uses this hour-long break to eat at his desk while catching up on the news in general, with particular interest in the physical asset management section. Darren is like a sponge; he likes to read about anything and everything that he can get his hands on. He especially likes to browse his favorite websites, one of them, of course, being triathlon-related. He likes to keep up-to-date with the different articles relating to training tips and to shop for new accessories and gadgets available on the market.

During lunch, Darren gets a call. It's Jerry. Darren picks up the phone, has a quick sip of water, and says in an excited voice:

"Hello young man! What's up?"

At the other end of the line in an equally excited voice, Jerry replies, "Doing well, Mr. Ordlan. How are you?" And almost instantly he continues, "Would you have a few minutes to spare?"

"I am doing well, Jerry. I am on my lunch break and do have a few minutes to chat."

"I really appreciate your taking my call during your break," Jerry responds.

"So let's get to business. What's up?" Darren asks again before putting a forkful of pasta in his mouth.

"Well, I read both Neel's paper and your article," Jerry starts in a calm tone. "They are both engaging and informative. I learned a lot regarding the physical asset management strategies and the components of physical asset management planning. Now I have a good idea of my next few steps at work to keep our asset management program moving forward. I wanted to discuss with you to see if there is anything else beyond what's in the documents that I should pay attention to."

"Jerry, I think you now have a good sense of all the major components of your physical asset management program," Darren starts. "However, you need to understand what it will take to implement them in your organizational context."

"Implementing asset management strategies and plans in an organization is a huge endeavor," Darren continues. "To realize maximum value from the physical assets, there are lots of strategies that need to be aligned, many activities to be coordinated, initiatives to be integrated, and plenty of collaboration to take place across the entire organization."

"Jerry, I don't want you to make the same mistakes that I made a few years ago," Darren says in a serious tone after a slight pause. "Back then I focused most of my and my team's efforts in developing frameworks, models, tools, and policies. I'm not saying these are not important. They are and are absolutely necessary. However, what I did wrong was neglecting several key components critical to successful implementation of a physical asset management program. They are called the 'enablers' that can make or break your program" (Figure 7.1).

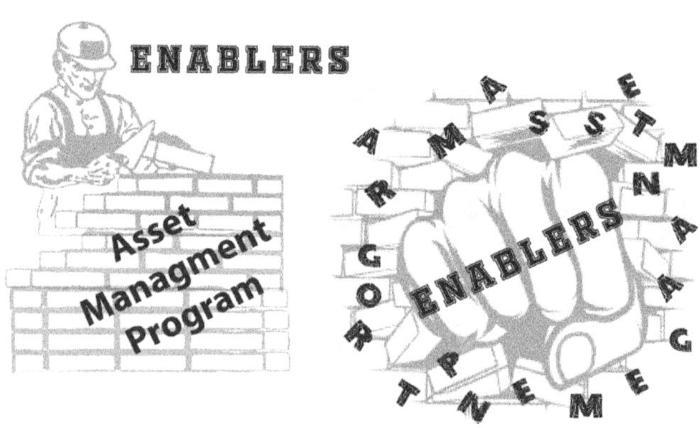

FIGURE 7.1 Enablers—make or break.

Data

"The first one," Darren says, "is data, of course. Organizations rely on a diverse array of physical assets to deliver the required levels of service. The purpose of physical asset management is to make the right decisions at the right time to minimize the total costs of acquiring, operating, maintaining, and renewing physical assets while maintaining the performance standards. Having the right information at the right time is essential to support decision makers to make optimal and timely decisions. This is why collecting and maintaining appropriate physical asset data over their whole lifecycle is essential to achieving those objectives."

"I think we do have some good data at Mouroum Inc.," Jerry interjects.

"I don't have any doubt about that, Jerry," Darren replies. "But I can first of all say with near certainty that the data you have is residing in different databases, and, second, that the data is not up-to-date. You may argue what is wrong with that. Long story short, the data is not complete and accurate enough to provide good information for you to use in the decision-making process."

"So what can be done about that?" Jerry asks. (See Figure 7.2.)

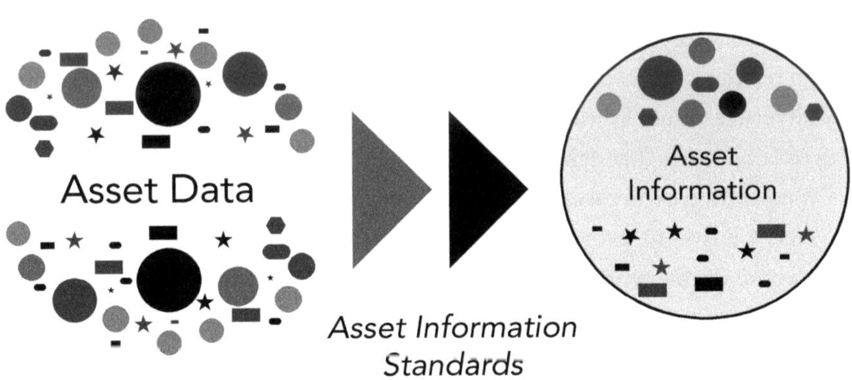

FIGURE 7.2 Data transformation process.

"There are a few things that need to be done right off the bat. First is to establish a data management strategy which will ensure that critical information needs and data requirements are defined. Then to document the location and use of the different data sources, information systems and tools out there. This will entail developing a data management model consisting of:

- Data quality standards for key physical asset data sets
- Data architecture for key physical asset groups in the form of physical asset hierarchies (which I am sure you have already)
- Physical asset registries to capture data attributes and fields in a standard manner"

Jerry is silent at the other end of the line, absorbing the information and taking notes as Darren continues to speak. "One more thing before we move on," he says, "is that you may also want to consider establishing a data governance model in your organization. The governance model will define data ownership, roles and responsibilities, and sharing of the key physical asset data sets. Gone are the days when physical asset data was treated in an ad hoc manner; today it is considered a critical element for an effective physical asset management system."

People

Darren continues, "While data is the cornerstone of an effective physical asset management system, the people are also a crucial element of the program. The essence of a successful implementation of physical asset management in an organization resides in its people. They are at the forefront of managing physical assets and delivering value. So they need to be engaged from the very beginning and involved at all times throughout the process.Ú

"Who should be involved?" Jerry finally asks after having listened for a while.

"Well it has to be everyone in the organization to some degree without forgetting some of the external players depending on your organizational context," Darren replies. "ISO 55000 does a good job in outlining the importance of understanding the organization's internal and external context and the expectations of stakeholders. In your case, Jerry, I would advise you to identify your organization's internal and external stakeholders such as key employees, senior management members, functional areas, owners, customers, suppliers, contractors, regulatory bodies, agencies, or local communities, etc."

Darren adds: "This will become very handy when you develop and roll out your organization's communication plan to ensure all stakeholders of the organization are kept informed at all times and on a regular basis regarding any update related to physical asset management. Implementation of an organization-wide physical asset management philosophy will have to engage staff at all levels of the organization to achieve key objectives such as:

- Creating awareness around physical asset management and explaining why it is important.
- Educating stakeholders on what physical asset management is about and how it impacts an organization.
- Keeping all stakeholders informed on how the physical asset management program is being implemented across the organization.
- Tracking and communicating progress of the implementation plan.
- Reporting on the successes of the physical asset management program to show the benefits."

"What are some of the methods used to achieve these objectives?" Jerry asks.

"There are various mechanisms you can use to roll out your communication strategy," Darren explains. "Most will not cost much to implement, such as brochures, newsletters, progress reports, or even departmental presentations."

"Maybe I can use our intranet for that purpose as well!" Jerry says suddenly.

"That's a great idea, and it will definitely work!" Darren quickly replies.

He then asks, "Do you have any other questions before we move on to the next enabler?"

"No, please carry on," Jerry says.

Information Technology

"The next enabler that you would want to focus on and that is a major issue for organizations is information technology," Darren states. "Here I would heavily stress the point of having fully functional physical asset management information systems and tools to support the previously mentioned two enablers, data and people. A fully functional physical asset management information system is a combination of hardware, software, and procedures and standards to enable people to collect, store, classify, and convert data into useful information. It has become an essential tool to manage physical assets in order to effectively deal with the extent of analysis required to support physical asset-intensive organizations and to handle complex planning and optimization scenarios."

"I agree with you, Darren," Jerry says. "Having a good information system in place will help a lot. But what I am seeing at Mouroum confuses me. First of all, the maintenance department is using an outdated system with unreliable data. I recently asked for a report on the lifecycle cost of a physical asset, and the department was not able to produce a complete report. On the other hand, I know the fleet manager is requesting a software system to manage its operations and maintenance."

"I understand your situation, Jerry," Darren responds. "This is very common in many organizations: either existing software systems are not used to their full potential, or you have multiple systems existing which are not connected to each other in any way. Both are a total waste of resources that do not bring value."

"Hmmm," Jerry ponders and then continues, "now that I think of it, I'll also need an information system to help me with good physical asset management planning. What would you recommend?"

"Jerry," Darren promptly replies, "I suggest that you first look at having a governance structure in place such as a steering committee to provide direction and control. Then you will have to do an internal and external scan of what you have in place in terms of information systems or data silos. This can be done either internally or by using external resources. It will help identify gaps and inefficiencies as well as identify potential opportunities. Once complete, the results can be reviewed by the governance team for further direction. More often than not, this will entail putting together a cross-functional and departmental team to document information requirements for physical assets and for key business functions, core and advanced system functionality requirements, integration requirements with existing network, long-term needs, and resource requirements and also to develop a corporate plan. Mind you that the whole process may result in some trade-offs and sacrifice, as some existing systems might have to be phased out and new systems brought in with different functionalities. At the end of the day, you want to ensure that the new physical asset management information system(s) is able to:

- Maintain a central up-to-date physical asset register with specific functionalities.
- Cost-effectively manage physical assets over their whole lifecycle.

- Provide efficient work management functionalities to respond to service requests.
- Manage resources and provide performance reports and performance measures.
- Capture, manage, and analyze physical assets transactional data throughout the whole lifecycle.
- Provide the platform for regulatory compliance and auditing purposes."

"This is a huge undertaking!" Jerry exclaims.

"Yes indeed," Darren says. "It is a huge undertaking but also the best way to come out of the data-rich, information-poor syndrome we are in. It will take some time to get there, but it is the inexorable need for information technology that will translate into building an effective physical asset management system."

"I should also mention one important caveat here," Darren quickly adds. "The organization and people need to be ready for change and be prepared to both commit and adapt to changes in business processes. This leads us to our next and last enabler—processes."

Processes

"As you can imagine, this is the toughest one. This is where everything we have talked about so far coalesces—the data, people, and technology converge into practical business processes and procedures. Let's face it," Darren comments, "business processes have always existed in organizations, and many, if not most of them, work well. Currently, organizations have processes to perform the required physical asset lifecycle delivery activities, but, in many instances, they lack:

1. Consistency across the organization as a management system
2. Alignment with the strategic physical asset management objectives
3. Alignment and integration with other management systems"

"Management systems," Darren continues, "are those processes needed for organizations to improve inefficiencies in their business activities. Similarly, for physical asset management an effective physical asset management system is required to improve those inefficiencies while also better managing risks."

"Why would you say that this is the toughest part then?" Jerry asks. "Well," Darren continues, "the conundrum now is that organizations are trying to build a different management system, a different way of doing business, a new paradigm to deliver best value from physical assets where people, data, and technology are heavily involved."

He explains, "In a typical physical asset-intensive organization, there are many lifecycle management activities taking place spanning across the whole lifecycle of physical assets. Those lifecycle management activities take place right from the planning, acquisition, installation, and commissioning stages through to operation, maintenance, and disposal. All the activities have various degrees of impact on the strategic physical asset management objectives. What makes the situation tougher is that those activities are performed by different people in different functional areas in the form of business processes and procedures. In my book, *Physical Asset Management: An Organizational Challenge*, I talk about the chaos and challenges that this can create. All the people in different functional areas are trying to do their best by performing in accordance with their processes and procedures. However, people doing their best is not good enough. First you need to know what to do and then do your best. This implies reviewing the existing business processes for the different lifecycle management activities to make sure they are interrelated. This also implies there is adequate interaction between the people in the different functional areas to ensure the strategic physical asset management objectives are met."

"How do we go about achieving that?" Jerry asks as he starts to understand the whole situation.

"As I mentioned earlier," Darren carries on, "this is the toughest part, as we are attempting to transform a business with its rooted culture of business procedures by building a new system with new and improved ways of doing things."

"Many organizations stumble at this stage," Darren adds in a serious voice.

"Why's that so?" asks Jerry.

"This is complicated to explain and will take some time," Darren replies. "It's already past one o'clock, and I have a meeting to prepare for. That reminds me—there's a one-day information seminar that I am also working on. I will be delivering it next month at the university to undergraduate and graduate students. The topic is on organization-wide physical asset management and will most probably touch on some of your questions about the business processes challenges in organizations. You may want to attend if you are available."

"Sure, that'll be interesting!" Jerry exclaims. "How much will it cost, and how do I register?"

"It's going to be a free event for the university students, and I plan to open it up to industry practitioners as well. I'll put your name down and will make sure you get the invitation once it goes out," Darren replies. "For now, I have to go. Nice talking to you. Take care."

Darren hangs up. Jerry feels content that he was able to connect with Darren and that he can follow up with him next month at the seminar. His mind already starts to reflect on the enablers, how important they are, and what he can do about them at Mouroum Inc.

CHAPTER 8

THE OPERATIONALIZATION CHALLENGES

It's 7:45 on a chilly Thursday morning, and Jerry drives into the parking lot of the University of Ryerson, a beautiful building right in the middle of downtown Toronto. This is where Darren is holding his seminar "Organization-wide Physical Asset Management," and Jerry doesn't want to be late.

He quickly finds a parking spot and speed-walks to the building. Jerry manages to find the lecture auditorium after having asked for directions along the way. Darren is standing at the doorstep and is getting ready to close the door while he greets the last few students entering the auditorium.

"Just in time, Jerry!" he exclaims. "Good to see you. Please have a seat anywhere you like."

There are only a few seats available in the front rows, and Jerry takes a seat right in front of Darren's desk and the screen.

Darren walks to the middle of the auditorium and looks up to see a fully packed room with around fifty people in attendance. The place is still noisy, with students chatting among themselves as usual.

"Good morning, everyone!" Darren starts after he switches on his clip-on microphone.

The auditorium instantly quiets down.

"Welcome to this seminar on organizational physical asset management," Darren continues. He then goes on to introduce himself to the class.

"Now let's see whom we have in the audience."

"How many postgraduates do we have?" Darren asks. A little less than half of the people in the auditorium raise their hands.

"Great, I can see that folks are seriously preparing to enter the physical asset management job market," and Darren cannot stop himself from expressing his excitement.

"How many undergraduates do we have in attendance?" he asks next. Seven hands go up.

"Not bad. It's good to see students interested in physical asset management early on," Darren states.

"And, finally, are there any practitioners from the industry in attendance?" Darren asks. One-third of the audience members raise their hands.

Darren is surprised to see so many practitioners in attendance. "Oh, wow. I didn't expect so many from the industries in attendance. This is good news, but, at the same time, it signals that there is a demand for more information and perhaps some guidance on physical asset management out there," he thinks to himself.

"Without further ado, let's get started," Darren announces excitedly. "And by the way, I want this session to be interactive. So please feel free to jump in at any time."

Darren smiles and continues: "There's not one day that goes by that we don't hear or read about physical asset management in the news. It has become a key concern for many organizations in both the private and public sectors. Many of them have been getting ready and are well on their way to implementing physical asset management practices. Plenty of different tools, approaches, principles, conceptual models, and standards, such as ISO 55000, are now available. They cover the whole breadth and depth of guidelines and activities required to manage physical assets."

Darren pauses and stares at the ceiling in a thoughtful manner. Then he says in a serious tone, "Now the big question is how well these physical asset management principles, standards, and practices have been used and implemented in organizations."

He pauses once more to check if the audience is with him and continues, "In other words, how well have we operationalized physical asset management across organizations up to now? And does everybody understand the term 'operationalizing'?"

There is no reaction from the audience.

Darren waits a few seconds and then goes on to explain, "The word 'operationalizing' means setting something up, so that it can be used across the whole organization."

"Now, can we say that physical asset management practices have been successfully set up and used across organizations? What do you think?" Darren asks and looks across the auditorium.

A young lady in a red pullover in the third row raises her hand and replies in a calm voice: "I believe some organizations have been successful and others are still struggling. Many have strategies, plans, policies, and procedures that are well developed and documented."

"I cannot agree more with you," Darren responds right away. Pacing across the auditorium, he continues: "But how well have these strategies, plans, policies, procedures, and processes been institutionalized and ingrained in the day-to-day business activities of organizations? Properly implementing them is of paramount importance for an organization to realize the benefits of physical asset management. But there are some underlying challenges."

"Why is that so?" Jerry asks with his hand up.

Darren stops and looks at him. "The answer lies in the strategy execution," Darren replies and then continues walking. He adds, "We all know that strategic thought and planning are very important, but we must recognize that execution of the strategy is crucial to achieving success and reaping results."

He walks toward the whiteboard, picks up a marker, and continues speaking: "Strategy execution is an important and hot topic in management today. In fact, it's rated as one of the most challenging issues faced by many company executives."

"Any guess as to what would be the estimated percentage of strategies that are not successfully implemented in organizations?" he asks. A few answers are volunteered by people in the audience, and all of them state numbers below 40%.

Darren smiles and turns to the whiteboard, where he scribbles 60% with an exclamation mark. "Yes," he says, confirming what he wrote, "it's estimated that more than 60% of strategies are not successfully implemented in organizations!"

The young lady in the red pullover, whose name is Karen, asks, "Why is that so? 60% seems to be quite a high rate."

"It is indeed surprisingly high," Darren replies. "It has to be understood that successful implementation of a strategy is not merely about implementing plans or strategies. It is more than that. It is about understanding what needs to be done to actually drive business results."

Darren continues: "In their book *Execution: The Discipline of Getting Things Done* (2002), Larry Bossidy and Ram Charan make the case for execution as a discipline or 'systematic way of exposing reality and acting on it.' They explain that strategy execution should be looked at as a system of three core processes."

Darren turns to the whiteboard and draws the following diagram (Figure 8.1.):

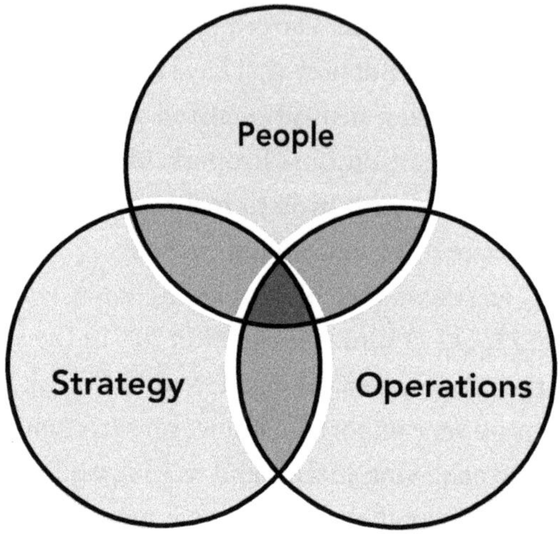

FIGURE 8.1 Three key processes of execution.

"People, strategy, and operations," Darren firmly says.

He pauses, stares at the ceiling, and then continues, "How these three core processes can be implemented by organizations to achieve strategy success is further explored by Professor Robert S. Kaplan and David P. Norton in their book *The Execution Premium: Linking Strategy to Operations for Competitive Advantage* (2008)."

Darren puts down the marker and continues pacing. And at the same time, he comments, "In the book the authors present their management system consisting of six sequential stages to help organizations measure and capture the increase in value derived from successful strategy execution."

Darren takes a deep breath and repeats aloud each of the six stages in the management system as if he has learned them by heart:

1. "Develop the strategy.
2. Plan the strategy.
3. Align the organization.

4. Plan operations.
5. Monitor and learn.
6. Test and adapt."

He then continues, "As you see, strategy execution is a step-by-step process."

"So where are organizations failing during the execution process?" a young man with glasses sitting at the back interjects in a husky voice, loud enough for Darren to hear him.

"Well, this is a tough question," Darren admits. "Organizations are all different, and so are the challenges they face. In my opinion, though, the stage that is most overlooked is Step 3 of the management system, 'aligning the organization.' It's perhaps one of the most critical aspects of the strategy execution process."

Darren walks over to his desk and takes a sip of water from his water bottle.

"To demonstrate this," he says, "let's go back to our earlier discussion about operationalizing physical asset management across organizations through strategy execution. How well are we executing physical asset management strategy across organizations, and what are the areas for improvement?"

He stops and looks at the audience as if waiting for someone to answer the question. Hearing no answers, Darren continues, "Let's take a closer look at the definition of physical asset management from the ISO 55000 series of standards."

He walks back to the whiteboard and writes the definition:

The set of coordinated activities that an organization uses to realize value from assets in the delivery of its outcomes or objectives.

"I want you to pay particular attention to three key words in this definition," he continues while underlining those words with the marker. And in a firm voice Darren repeats each of the underlined word:

1. "Activities
2. Coordinated
3. Value"

"Now, let's spend some time analyzing those key words in that order," Darren adds.

"First off," he says, "let's talk about *activities*. In organizations there are all kinds of daily activities taking place to keep the business running. Activities directly or indirectly related to physical assets are managed in different functional areas. They can range from simple asset maintenance routines to major asset design decisions or simple depreciation calculation exercises."

"Can anybody give some examples of those activities?" Darren asks, as he makes his way over to the flipchart and grabs a marker.

Darren can hardly keep up with the many examples that the audience members are giving—the suggestions coming mainly from the industry asset management practitioners.

After a few minutes of scribbling, Darren raises his hands and says in a loud voice: "OK. I think we have enough examples, and we can stop here! And we do know that the list can go on and on."

He steps aside to let the audience have a good view of the list on the flipchart.

These are the example activities that are listed:

- Updating the physical asset registry
- Identifying and prioritizing capital projects
- Developing the ten-year capital forecast
- Developing maintenance and reliability tactics for physical assets
- Carrying out condition assessments on all physical assets
- Designing a new production line with new physical assets
- Executing refurbishment work during a major shutdown
- Carrying out condition assessments
- Entering data in a work management system
- Compiling physical assets lifecycle costs
- Forecasting future needs and demands for physical assets
- Measuring performance of assets and asset systems
- Operating physical assets
- Developing and implementing asset management policies and procedures

"Is everybody satisfied with these examples?" Darren asks after a few minutes of silence.

He sees many heads nodding across the auditorium. "So far it looks like everybody is on board," Darren says to himself.

"As mentioned, this is only a partial list of physical asset management activities," Darren continues. "I suggest you review the more detailed and comprehensive list of physical asset management activities existing in an organization to understand which ones are relevant to your specific organizational context. Once you have done that, I suggest you ask yourselves a few questions."

Darren sits down behind his desk and starts posing questions, saying, "Try these to get you started."

"How well are the physical asset management activities being executed in the organization?"

"What functional areas are responsible for doing them, and is there real ownership and accountability around those activities?"

"Are there activities that are not done at all but are required?"

"And for my final question," he says, "Those activities that are done—are they done in a consistent manner, i.e., following standard policies, procedures, and processes?"

"Do you mean that these physical asset management activities *are not* being executed in an organization?" a young woman elegantly dressed in a blazer asks defiantly.

Darren looks at her and smiles. "No, I never said that," he proclaims. "Those activities may already be executed in many organizations. The point I am trying to make is, how well are they coordinated?"

"Which brings us to the second underlined word, *coordinated*, in the physical asset management definition," Darren adds.

Still sitting at his desk, Darren sinks further down in his chair and stares blankly at the audience.

"In my opinion, coordinating the physical asset management activities across an organization is one of the toughest challenges. No wonder why the success of strategy execution depends heavily on coordination of those activities in an organization," Darren confidently adds.

"What does coordination mean, and why is it such a challenge?" Jerry asks as he raises his hand.

"By definition, coordination means 'the organization of the different elements of a complex body or activity so as to enable them to work together effectively,'" Darren explains and then pauses.

He then continues, "In an organization this translates into the synchronization and integration of activities to ensure that the resources are used most effectively and efficiently to achieve the strategic goals and objectives. Coordination is one of the key functions of organizational management when it comes to identifying leadership, responsibilities, accountabilities, execution, and control of those activities. This is where people from the different functional areas will have to get involved and collaborate to develop, own, and execute the activities that are required to realize value."

Darren pauses again briefly to ensure the audience has the time to absorb what he just said. He then continues in the same tone, "Similarly, in physical asset management, a significant number of activities are identified, and are all very important to realize value from physical assets over their entire lifecycle phases."

Pointing to the flipchart Darren explains, "These activities are in the form of processes, practices, and business procedures and need to be well coordinated and integrated in the organization."

Everybody in the auditorium is now quiet. Finally, the young guy with glasses, Nihom, breaks the silence. "How does an organization achieve coordination?" he asks in his husky voice.

Darren gets up from his chair and starts pacing again. And as he does, he begins to speak: "First of all, organizations need to be properly structured with the required functional areas and adequately resourced with the right leadership and staffing. By adequately I mean personnel with the right expertise and skill sets who understand the importance of the different physical asset lifecycle phases and the technicalities involved in carrying out the physical asset management activities. Clear roles and responsibilities are also very important to ensure everyone knows who is responsible and accountable for which activities. This would support cross-functional collaboration to enhance management of business processes across the traditional boundaries of functional areas."

Darren takes a deep breath and continues: "This, in turn, will translate into coordinating and synergizing the activities of different functional areas to realize the superordinate cross-functional goals and objectives."

And, in a poised voice, he declares: "The onus is truly on the different functional areas of the organization and their personnel; how the functional areas individually approach the development and implementation of physical asset management activities; how they collaborate with other functional areas in the process; how value is created through them and carried over to other functional areas to be further maximized; how knowledge and information are shared cross-functionally to achieve value optimization."

Darren pauses, and with both his hands raised, he loudly proclaims: "This is the essence of coordination in physical asset management!"

A slight commotion can be felt in the audience with a few words being exchanged.

Darren waits a few minutes until the noise in the auditorium settles down. He then carries on: "To you industry practitioners, this will be your biggest challenge in physical asset management. And to you students, remember that coordination is important no matter in what functional area you end up dedicating your career."

Darren checks the time on his watch and continues: "I think we have just enough time to talk about the last of the key words in the physical asset management definition, *value*, before we take a break. So let's get on with it!"

Darren takes another sip of water, clears his throat, and sits on the edge of his desk. Then in a serious tone, he says, "In organizations *value* is the main focus, and it is befitting to spend some time understanding what value is and how it can be realized."

The auditorium is so quiet you could hear a pin drop.

Darren continues in the same tone: "Organizations design, build, own, operate, and maintain physical assets to obtain value from them over their full economic life. Value can be in the form of goods or services, tangible or intangible, financial or nonfinancial, depending on the organizational objectives, the nature and purpose of the organization, and the needs and expectations of its stakeholders. Understanding what value is can only help organizations in realizing value from physical assets. It will help in understanding what we are doing or have to do to *realize value* from the physical assets."

Darren notices one hand rise at the corner of the auditorium and pauses.

A young man asks in a soft voice: "Sir, how does an organization define value?"

"This is an excellent question," Darren promptly replies as if he were expecting it. "Unfortunately, I cannot give you a concrete answer, as value will differ from organization to organization depending on its operational context. Try to ask anyone in any organization what value means. You may be surprised to hear that either they don't know how to define it, or if they do know, the answer you get back may differ from person to person or from department to department."

"So how do organizations cope with that?" the young man asks, visibly puzzled.

"Well, we know that organizations invest a lot in adding value to whatever they do. They invest in the best technologies, invest in the best people, provide the best tools and training, and adopt effective practices to achieve that," Darren starts to explain.

He takes a breath and looks out at the audience. Then he raises his forefinger and continues: "But there is one major aspect that is frequently overlooked in organizations: How to stop destroying value."

"What? Destroying value? What do you mean by that? Nobody destroys value!" the young man, whose name is Zach, says in an irritated voice.

"I fully understand your shock," says Darren calmly. "I would also like to think that nobody destroys value in an organization, but unfortunately it does happen. And I should point out that most of the time it is unintentional."

"Can you please explain how this can happen? I would like to relate it to my workplace," Zach says in a worried tone.

"No worries," Darren continues calmly. "Destroying value happens all the time and will keep on happening unless we do things differently."

"What do you mean by doing things differently?" Jerry interjects.

"It looks like everyone is getting confused," Darren says to himself. "I better provide a concrete example to clarify."

Darren stares at the ceiling, his usual source for inspiration. The whole auditorium goes silent again waiting for his response.

He takes a deep breath and resumes pacing across the auditorium.

"Let's take an example you are all familiar with," Darren starts. "I'm sure you all have some plans in your personal and professional lives to achieve certain goals. I'm also sure you will have put together a to-do list to help you achieve those objectives. Am I right?"

Darren sees many heads nodding in agreement, and he continues, "Have you ever put together a 'not-to-do' or 'stop-to-do' list?"

There is total silence again in the auditorium for several seconds.

Finally, Jerry breaks the silence. "Well, not really," he quietly replies. "I somehow know what not to do or what to stop doing, but I have never written anything down."

"Thank you for your honest answer," Darren says smiling at Jerry. "Would it be correct for me to say that you do not focus too much on the not-to-do things and rather you focus mainly on the to-do list?" He stops and looks at Jerry.

"Yes, that's somewhat correct," Jerry reluctantly replies.

Darren turns around and bangs his fist on his desk. "This is exactly the situation prevailing in many organizations!" Darren says in a sharp and convincing voice. "They focus mainly on how to create and add value, but not enough on how to stop destroying value."

Karen raises her hand and asks in her soft voice, "Can you please elaborate on that from an organization's perspective?" She adds, "With an example if possible."

"Absolutely, let me explain," Darren excitedly says. He seems to enjoy the questions thrown at him, as they force him to think more on the topic at hand.

He takes another sip from his water bottle, quickly glances at his watch, and starts: "Let's say that our organization produces fresh bakery products. We invest in the latest and greatest technology to have a state-of-the-art production line producing high-quality products to meet the demands of our customers. We hire qualified technicians and train them to be proficient at troubleshooting and repairing equipment. On the other hand, our procurement department has mandates to buy parts from specific suppliers to take advantage of economies of scale and maintain low costs. This leads to substandard parts being used on the high-tech equipment and resulting in more frequent equipment failures."

Darren pauses to reflect on what he just said. Then continuing on, he says: "So on one side, the organization is trying to create and add value to meet customer demand by running high-performing equipment with highly trained personnel, while on the other side within the same organization, value is being destroyed by having unreliable equipment because of substandard parts."

"Oh, I see," Karen says while nodding her head, as if the puzzle pieces are starting to fit together.

"This is only one simple example. There are many such examples related to physical asset management activities happening daily in lots of organizations," Darren adds and then takes a pause.

"I have to say though, many of those value-destroying activities, in most cases, are unintentional," Darren adds thoughtfully. "It's just the way organizations are structured and managed."

Darren looks at his watch and says, "Before we take a break, let me further elaborate on the topic of value. First and foremost, we must understand, define, and communicate the meaning of value in an organization."

"What is value, and how does an organization define the term?" Darren then asks.

He scans the audience for answers. After a few moments of hesitation, Zach raises his hand and replies, "It's what customers expect to get from products or services."

"Not only customers but also the shareholders and other stakeholders of the organization," Jerry argues vehemently.

"It can also be the physical asset performance and contribution to achieve organizational objectives," Karen adds.

Darren is happy with what he is hearing. "All absolutely correct," he says. "It's critical for organizations to define those values from different angles, document how they are realized, and determine how they can be quantified and measured based on the different organizational and operating contexts."

"These days we are seeing new trends within organizations. They are being pushed to dial down their single-minded pursuit of financial gain and pay closer attention to their impact on society. Successful organizations are those that have a good balance to commit to economic, environmental, and social values and build them into their corporate strategy."

"Now, to realize those values, an organization has to execute its strategy properly to maximize return on the physical assets invested in. These days many organizations are struggling with strategy implementation, as I mentioned earlier. They struggle to realize value from initiatives and programs and to deliver positive sustainable results on a consistent basis."

Jerry looks at Darren anxiously. "Why is that so?" Jerry asks. "Most organizations have the required physical assets in operations, the trained personnel in place, and fully developed strategies lined up."

Darren looks at him, and with a grin on his face, he continues, "As I mentioned earlier, it all has to do with organizations."

He pauses and takes a look at his watch and then announces in a cheerful voice, "Guess what? It's time to take a well-deserved break! Let's reconvene in fifteen minutes."

CHAPTER 9

ORGANIZATION—A COMPLEX SYSTEM

ORGANIZATION

The auditorium is almost full again with only a few empty seats. Fifteen minutes have elapsed, and Darren wants to start right away, as he thinks about the packed agenda he has for the rest of the day.

"What is an organization?" he starts. "This may sound like a weird question to ask, but I think it's important to start with the basics."

He takes a pause as he watches a few more students take their seats. Then ready to go on, he says, "There is no agreed-upon definition of an organization, but the key concept is that it's a group of people, working together, pursuing common goals or objectives. Over time the focus has shifted to other aspects of organizations, including the structure and the way in which work is organized, the people and their needs and relationships with others in the organization, the ways in which organizations interact with their external environments, the organizational culture, or the values and beliefs prevailing."

The auditorium has gone quiet again, and everybody seems to be focused on what Darren is saying.

He continues: "Basically, an organization is a social unit of people that is structured and managed to meet a need or to pursue collective goals. Every organization has some sort of management structure that determines relationships among the different members and their respective activities and subdivides and assigns roles, responsibilities, and authority to execute the tasks at hand."

He pauses and points at a diagram he drew during the break on the whiteboard, specifying "This is a simple model of an organization" (Figure 9.1.).

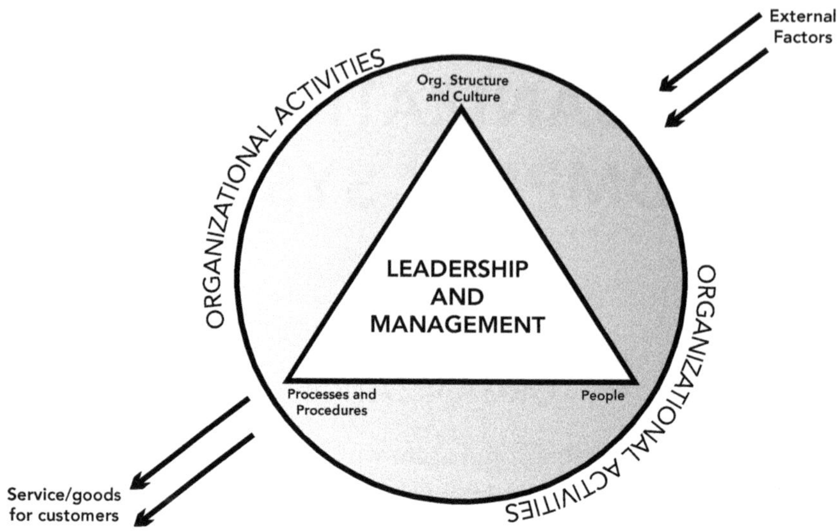

FIGURE 9.1 Simple model of an organization.

He directs his laser pointer at the words "Org. Structure" and notes: "An organizational structure is essentially concerned with the way in which work is divided among different individuals and different levels of the organization to meet the strategic goals and objectives. The structure also allows the work of individuals and groups to be supervised and coordinated at different levels."

He then shifts his pointer to "People," saying: "People are key for organizations, as they are the prime resource to get the work done. People work in groups or individually, arranged within the boundary of the structure. In organizations it is important to keep people motivated and committed to work. This is where management and leadership become very important."

Everybody's eyes are now riveted on the whiteboard while paying attention to what Darren is saying.

He then points to "Processes and Procedures" and adds, "Organizations have a number of different processes, subprocesses, and sets of associated procedures through which they operate depending on the type of organization and the business context. These processes and procedures are executed and managed by the people of the organization within the limits of the organizational structure."

He puts down his pointer, and clasps his hands together, and faces the audience. "Depending on the nature of their business," he tells his listeners, "organizations can become very complex. That same complexity is their biggest challenge. To deal with complexity, organizations regularly review their strategy and structure to try to bring about improvement in line with the changes going on around them and in their external environment. Some organizations bring on corporate-level functional managers, who, with the help of teams of professionals and staff, try to exploit potential synergies and to coordinate activities to deliver results. But there are many frustrations and failures along the way, as things are more challenging than they look."

He moves to his desk and picks up a book, waves it in the air, and says: "In my book, *Physical Asset Management: An Organizational Challenge,* I delve a bit into those challenges that organizations are facing when implementing a holistic physical asset management philosophy."

Zach raises his hand and asks, "I have not read the book. Can you please explain what those challenges are?"

"Zach, you definitely have to read the book. It'll take too long for me to go through all those challenges here," Darren replies. "But what I can tell you is that organizational culture and leadership are two of the key sources of frustrations and failures that impede successful strategy execution in organizations. They are actually the foundation for any organization striving to achieve excellence, and, as we well know, not many organizations have been able to make the leap from good to excellent."

Darren puts the book down and starts pacing.

"Let's take 'organizational culture,' for example," he says looking down. "An organization's culture may be related to the way that the organization deals with its clients or customers, the way that staff interact with each other, and the nature of relationships between managers and their staff. It is the dominant pattern of shared beliefs and values that prevail in an organization."

He pauses and looks at the audience for any reaction. He sees none and continues, "Very often the style of leadership within an organization is a major contributor to its culture."

Darren could hear a few murmurs in the auditorium and smiles. "Let me explain," he adds. "A lot has been and is still being written on the topic of leadership. We are not going to get into the details of leadership and leadership styles, but there is one important thing to understand when it comes to organizations: the role played by 'leadership.'"

Darren stops right in front of Jerry and looks at him, saying, "In organizations the role of leadership is to build and motivate a team to work together collaboratively to transform the organization and achieve its strategic goals and objectives. Good leadership entails different key traits such as business acumen, people skills, decision-making skills, strategic vision, innovation, and credibility, among others."

He turns around and starts pacing again.

"What am I talking about here?" he questions aloud and takes a brief pause. Then continuing with his train of thought, he adds, "Many organizations fail in executing their strategy due to poor leadership during the transformation process, but, on the other hand, many organizations have succeeded in the path for transformation thanks to their strong leadership."

Jerry puts up his hand and asks, "So, what should organizations do about culture and leadership?"

"Great question, Jerry," Darren replies. "Unfortunately, there is no off-the-shelf solution for these challenges mainly for two reasons."

"First," he continues, "all organizations are different. From private to public sectors, to different industries and different complexities. Each organization will have its own set of issues and challenges based on the particular context. Second, in organizations we are dealing with people at different levels with different needs and beliefs. It is a huge challenge to create long-lasting synergies among people right across the whole organization."

"To try to answer your question, Jerry," Darren says as he looks back at Jerry, "one key factor that could be detrimental to resolving or at least alleviating these challenges is the organizational structure. Organizational structure is the way organizations are designed and arranged in a set of hierarchies to delegate responsibilities in order to achieve their goals and objectives. However, the concept of organizational hierarchy is no longer adequate and effective. There is something that is preventing organizations from facing and navigating the prevailing realities of complexity within their organizations. Traditional hierarchical/functional models still exist at different levels today, and yet there are no signs of shifting toward a better solution."

"Why is there reluctance to shift to something better?" Jerry inquires.

"Well, the concept of organizational hierarchy has been around for quite awhile and is embedded in our mind as a standard, natural way to build an organization. The assumption here is that by breaking a big situation into smaller pieces, more control can be exerted over every piece. That was the original premise."

Darren takes a deep breath and then adds in a forceful manner, "Although this approach effectively delegates responsibility in a hierarchical way, it unfortunately also creates artificial walls and ceilings resulting in functional silos."

He again pauses and then continues in a disappointed voice: "Unfortunately, the mechanistic worldview that believes that the results of the whole organization equal the sum of its individual, separate, and hierarchical parts does not hold anymore."

The auditorium again turns into total silence as if the audience is mulling over Darren's last statement.

"Sir, you mentioned 'functional silos' earlier," Karen finally breaks the heavy silence in the auditorium. "I read about it in some literature, and it is a common issue. Can you please elaborate on how it impacts organizations, and particularly physical asset management, with an example?"

Functional Silos

Darren clears his throat, folds his arms, and sits on the edge of his desk.

"Everyone hates silos, but they are natural and unavoidable," he starts. "As organizations develop deep areas of expertise, functional areas and departments inevitably start to find it more difficult to collaborate and work together. They don't have the same goals, and they have different mandates. Symptoms of functional silos in organizations are not that difficult to identify."

"There are many symptoms affecting individuals as well as functional areas or departments across organizations. Some of the very common ones are," and he starts listing them aloud:

- "Functional areas or departments focusing on their own set of objectives
- Poor communications taking place between functional areas and departments
- Lack of collaboration across functional areas due to poor understanding of what each one does
- Poor level of innovation due to lack of trust and knowledge sharing
- Projects and programs that are poorly executed and delayed
- Poor working atmosphere and employee morale resulting in frustration and resentment"

Then he continues, "The effects of functional silos are very much widespread in organizations and have significant implications. Individuals or functional areas cannot see or do not want to see the bigger picture. This is because either they may have been blindsided by everything else happening around them, or they have been instructed to operate that way. There are so many causes of the emergence and institutionalization of functional silos in organizations as explained in my book."

"It looks like I definitely need to read that book," Zach quickly comments.

"It can be a good thing if you want to catch up with my train of thought," Darren replies with a smile.

"Going back to Karen's question," Darren says, "to illustrate the magnitude and effect of functional silos across an organization, let's say that the director of operations makes the decision to cut 5% from the maintenance budget."

Darren looks around the auditorium in anticipation of some reaction.

Nihom, who has been quiet for some time, raises his hand. "That's a normal occurrence in many organizations!" he says loudly in an excited voice. "It happens in my organization too."

"Yes, this is a common thing," Darren continues. "It happens all the time, but unfortunately the effect of this decision is not as fully assessed as it should be. This is mainly due to the functional silo syndrome existing, where the focus is on one's own set of objectives. Eventually a moment will come when the organization will feel the effect of that decision, such as recurring physical asset breakdown and low availability meaning low throughput. Nobody will ever know (maybe it will take some time) or nobody will ever admit that the problem originated in the arbitrary reduction of the maintenance budget."

"I fully agree," Jerry quickly admits. "I cannot see anyone coming forward to admit that. Most of the time, the blame will be put on someone or something else."

Jerry sounds a little frustrated, and Darren can feel it in his voice. He responds with understanding by saying, "These are unfortunately some of the effects of a traditional organization and its functional silos, where all too often the focus becomes power games and personal success as opposed to achieving the goal of the organization."

"Wait a second!" Nihom exclaims. "Is this not an example of destroying value?"

"You are absolutely right!" Darren bursts out with excitement. "I like the way you are connecting the dots."

Then in a lower tone he continues: "It is unfortunate that even though in some parts of the organizations great things are happening to add value, a simple decision somewhere else in the organization can actually destroy that value and make things worse. Isn't that frustrating?"

Darren again waits for some reaction from the audience.

"Of course, it is very frustrating," says a young fellow in a blue shirt sitting next to Nihom. "I go through that all the time!" he says, irritated. "You know what you have to do, and you try to do your best, but, in the end, you run into roadblocks with other functional areas that are reluctant to help."

Darren looks at him and tries his best to calm the situation down by steering the discussion in another direction. "I feel your frustration," he says. "So, let's analyze what is going on from an organizational perspective."

Darren points again at the diagram on the whiteboard, and as he does, he says: "One of the biggest challenges organizations are facing is complexity, and with time, the challenge worsens with increasing complexity. In the twenty-first century, every organization needs to be prepared to maneuver through complexity. This applies to any organization whether in the public or private sector and in any industry. The problem is that organizations are still very much trapped in an outdated paradigm of functional silos, fragmentation, and command and control structures as reflected in the traditional organizational hierarchy mentality. This mechanistic and segmented approach was successful years ago, but with the inherent nature of modern organizations, it does not work anymore. Organizations need to reinvent and transform themselves to be able to navigate the challenge of complexity."

He takes a slight pause to allow the audience time to absorb what he just said. He then adds, while weighing heavily on each word, "Organizations need to undergo a paradigm shift to move toward an urgently needed systemic understanding to promote coordination and collaboration across all functional areas."

As soon as he finishes the sentence, Karen puts up her hand with a question. "What do you mean by 'systemic understanding'?" she asks in a quiet voice.

Darren was actually expecting this question, which aligns exactly with his plan to steer the direction of the whole discussion. Without delay he replies, "Systemic understanding means to provide a holistic approach to understanding the problems and challenges relating to a system, as opposed to a particular part."

"A system?" Karen looks confused. "What does 'system' have to do with physical asset management?"

"A lot," Darren replies. "System has a lot to do with physical asset management. I will explain that later, but for now let's dive into what a 'system' is about and understand its complexity."

System

Darren goes back to his desk and fires up the projector, which is on standby. He then opens a document on his laptop and projects it on the screen.

"Take your time and read through the following definitions," he tells the audience.

While everyone is reading the screen, Darren sits down at his desk and checks his phone for any messages or missed calls. He drinks some water from his bottle and starts reading the document on his laptop:

> A system can be defined as a network of either natural or manmade interrelated components, which interact cohesively together to achieve the goal of the system. General systems theory has been around for many years dealing with broadly applicable concepts and principles and, over time, has evolved into the interdisciplinary study of sys-

> tems. The goal of systems theory is systematically identifying a system's dynamics, constraints, conditions, and developing principles that can be applied to achieve the system's optimization. The components that make up a system are linked together by a relation of interrelationships and interactions to create a network of interdependencies. Interrelation, interaction, and interdependence are three distinctly different phenomena.
>
> *Interrelation* is the way in which each of two or more elements is related to the other or others. In a system a set of elements may not be related directly to one another, while another set may have a high degree of interrelationship. If two or more elements of a system are highly interrelated, there must be a significant level of interaction between those elements to enhance the interdependencies between them. Before there can be any form of interdependence, there must be interaction between the interrelated elements of the system. *Interaction* defines an exchange of some kind between interrelated elements of a system. As interaction increases, this presents more channels or mediums for the development of interdependencies within the system. *Interdependence* is a type of connection or relation between elements and is one of the central concepts within systems theory. In systems theory, interdependence means an interrelationship between autonomous elements through which interactions take place resulting in the emergence of an overall system that is greater than any of the parts.
>
> Although they are different ideas, interrelation is a prerequisite to interdependence, and interaction is a driver of interdependency. There may be other factors that change the degree of interdependency within a system, but interaction is a primary factor.

After having read the definition, Darren waits another couple of minutes. Then he stands up and starts pacing across the auditorium.

"So," he utters, "any questions or comments?"

Karen raises her hand right away as if she were waiting for the moment. "I still cannot see how all this ties in with physical asset management," she says with an impatient voice.

"Wait a minute," Darren replies with a big smile. He really appreciates Karen's eagerness to get to the point, but deep inside he feels he should explain the logic behind it. "We're going to get to it in just a moment," he adds. "First let's go back to organizations and apply the concept of systemic thinking."

Darren stops right in the middle of the auditorium, faces the audience, clasps his hands, and starts: "Systemic thinking is the ability to see things as a whole (or holistically) including the many different types of interrelationships, interactions, and interdependencies among the many elements in a complex system. The systemic approach to management in organizations is based on the general systems theory—the theory that says that to understand fully the operation of an entity, the entity must be viewed as a system. Earlier you read that the goal of systems theory is to systematically identify a system's dynamics, constraints, conditions, and developing principles that can be applied to achieve the system's optimization."

"Organizations are inherently systems made up of complex interrelations, interactions, and unprecedented levels of interdependencies among its functional areas, creating an integrated whole that is much more than the sum of its parts. The properties of this whole derive from the existing interrelations, interactions, and interdependencies, which in turn creates the true context of an organization. It involves the emergence of some overall combined system to the organization where individual functional areas retain their autonomy with respect to each other, but they are interrelated, and interdependent, and they interact with respect to the overall combined organization."

Darren pauses and starts pacing again. It looks like he gets his inspiration by walking around.

He continues speaking: "Functional areas differentiate their functions and status with regard to other functional areas in order to create a combined overall system. It's this relationship of interdependence that shapes the complex organizational systems. New dynamics emerge within those systems as a whole—dynamics which do not exist when considering functional silos separately. Imagine for a moment a financial department or a purchasing unit operating in a vacuum on its own. There is no purpose and no essence for its existence."

"The change in the parts is not seen to be generated by any elementary part but instead is defined by its interdependence with the whole. There is more cooperation, communication, and interactions among individuals, functional activities, functional areas, and departments, creating synergies and the emergence of a new level of organization. This emergent organization is not associated with the individual properties of any of its parts; thus, no part is directly dependent on any other. They are shaped and formed by the whole organization. These new dynamics define the essence of an organization, where dividing things up is no longer the answer. These same dynamics, if properly coordinated, are the ones that influence success and realize value in organizations."

He walks up to his laptop and pulls up a slide with a diagram on the screen (Figure 9.2). "This is how I would illustrate what I have just said."

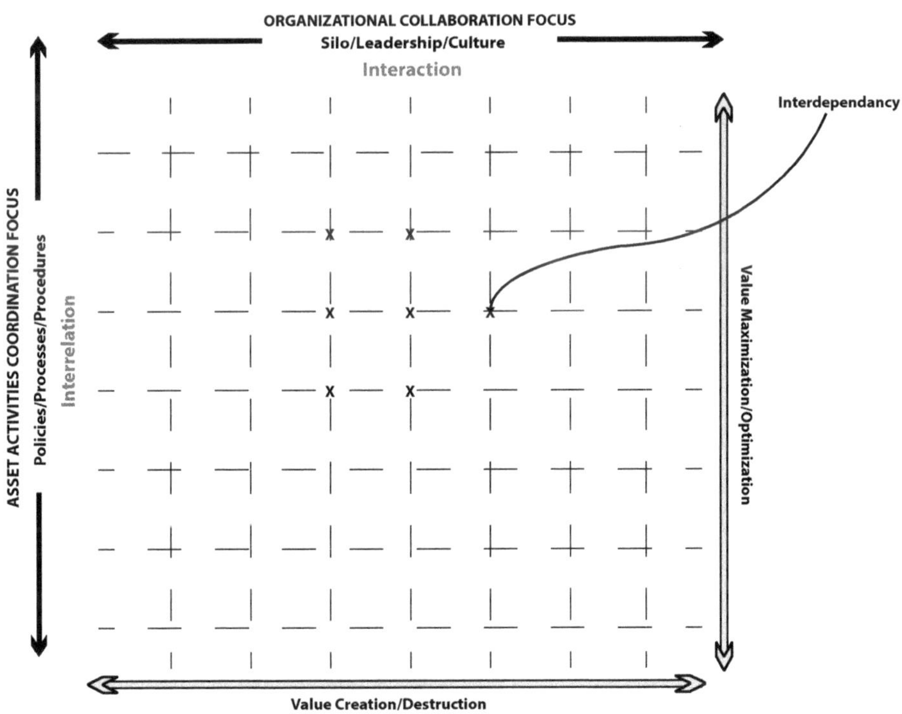

FIGURE 9.2 The organization system.

Darren walks away from the screen and lets the audience studyrtfgt the diagram and assimilate its content. Then he turns around and points at the diagram on the screen and explains: "The horizontal axis depicts the focus on interactions in organizations to create value in physical asset management and not destroy value. The focus is on bridging the silos and instilling strong leadership to foster a culture of collaboration."

"The vertical axis illustrates the need for interrelations across all the physical asset management activities happening over each phase of the entire lifecycle. The focus is on policies, processes, and procedures that need to be coordinated to realize value from physical asset management activities."

"Interactions and interrelations of physical asset management activities in different functional areas form a complex network of interdependencies. Those interdependencies take the form of some types of physical asset management activities performed by certain functional areas occurring at different levels of the organization."

Darren then points his laser pointer at the "x" intersections on the diagram and says: "These are the interdependencies between interactions and interrelations that make up the complexity of a system. This is where the focus must be!"

The auditorium goes into another heavy silence as people are trying to wrap their heads around what Darren just said.

Finally, as he often has before, Jerry breaks the silence, asking,"What is forbidding organizations to focus on those interdependencies and coordinate those dynamics to realize value?"

Darren closes his eyes and goes into a few seconds of silence as if he is meditating. The whole auditorium waits eagerly for what will happen next.

Darren finally opens his eyes, and with a large smile he starts: "I would cite the book *Quality, Involvement, Flow—The Systemic Organization*, by Lepore, Montgomery, and Siepe (2016). The authors assert that there are three major concerns that every organization today should focus on in order to deliver success: quality, involvement, and flow."

He pauses for another moment and then continues. "As defined in the book, '*quality* is grounded in the belief that management in any shape or form is prediction, and the ability to predict is the essence of any epistemology, or theory of knowledge. Quality is what allows an organization to differentiate itself from its competitors. Sadly, in a silo-dominated world, quality is often relegated to a department. *Flow* is the result of our actions—flow of materials, information, and money. Flow is associated with motion, and hence with evolution and the pace of change. It is the derivative of space and time, and it reminds us that no real value can be created unless a certain amount of change is taken into consideration. *Involvement* of people is what enables quality and flow. The foundation for any involvement is the teamwork that underpins the functioning of organizational systems. In a system, we can ask everyone to subordinate to the goal of that system. A correctly designed system can resolve conflicts of self-fulfillment and subordination to the goal by allowing an individual to gain more by subordinating to the goal than he or she would achieve independently.'"

Darren stops and takes a deep breath. Then he continues: "I hope I got it close enough. Let me elaborate a little bit more on what I just said."

"Organizations, by nature of their work, are cooperative and cross-functional, with a system of recurring and nonrecurring activities. However, the current traditional hierarchical/functional structure of organizations does not support that and more often is a source of inherent conflict focused on power and control. To navigate through the network of interdependencies, to manage complexity, and to break the hierarchical conflict existing in organizations, a new kind of organization must be built—a new system with focus on coordination instead of functional silos, a new system that will create quality, flow, and involvement through interrelation, interaction and interdependency."

Darren stops for a moment, and looks at Jerry, and says: "To answer your question on what is forbidding organizations to focus on the network of interdependencies and coordinate them to realize value—this will require organizations to adopt a cross-functional management system. A cross functional systems approach to management is about system thinking and management of business processes across the traditional boundaries of the functional areas."

Darren looks at his watch and adds, "The cross-functional management system must have two key attributes: one is leadership to clarify its goals and objectives and deploy them to all employees at every level, and the other is a system of close collaboration and coordination among different functional areas."

"Time now to take a break and have some lunch!" Darren announces while rubbing his hands together. "You will have some time as well to reflect on what we have covered up to now. See you all back here in an hour."

CHAPTER 10

THE NEED FOR AN EFFECTIVE PHYSICAL ASSET MANAGEMENT SYSTEM

Darren has his lunch in his office while he catches up with his email. Today he's having his preferred turkey sandwich followed by a bowl of yogurt mixed with blueberries and almonds. After eating his lunch Darren still has twenty-five minutes before the restart of the seminar. He closes his eyes and starts thinking about what he will cover in the second half of the day. So far he likes the keenness and interaction of the audience. He is so glad that he has invited practitioners to attend, as they generate lots of great questions based on practical experience and challenges. This can only benefit the students in attendance.

Suddenly he opens his eyes in a panic to realize that he dozed off. He looks at his watch and feels reassured that he has slept for only about ten minutes. He still has another fifteen minutes to get ready. Darren quickly makes copies of a two-page handout and then gathers his belongings before making his way to the auditorium.

The room is still empty, with only a few students busy on their phones. He places a few copies of the handout on each row of seats and heads back to his desk waiting for the one-hour lunch time to elapse.

Time is up, and most have returned.

"Welcome back, everyone," Darren starts. "I left copies of a handout in each row. Please pass them around and read the first page."

First page of the handout:

Managing physical assets, by nature, is a cooperative and cross-functional activity which demands a solid governance structure, strong leadership, and well-coordinated policies and practices. Physical asset management is inherently a system of recurring and nonrecurring physical asset management activities happening at different stages of physical asset lifecycle phases across the entire organization. It is befitting to say that a cross-functional approach based on system thinking is required to effectively manage physical assets across an organization.

There are many misunderstandings about physical assets and physical asset management concepts and principles in different industries. It is, however, very clear that the profile of physical asset management in both the public and private sectors has increased and will continue to increase as we continue to face challenges in the form of aging physical assets, significant infrastructure deficit, tough economic situations, and climate change vulnerability. It therefore makes total sense that physical assets have to be managed effectively; otherwise the situation could get worse. To complicate matters, there is a vast array of different types of physical assets to deal with in organizations, ranging from discrete to linear and with different lifecycle time frames ranging from a few years to sometimes a hundred years. On top of that, organizations with different organizational structures and various functional areas deal with physical assets over their whole lifecycle. All of this forms a complex system with an overwhelming network of interdependencies. It is logical to assume, therefore, that the overall situation facing organizations is often highly complicated and highly variable. It follows that, to navigate through this complex situation and manage the variables, there is a need for an effective physical asset management system that is acceptable to the organization and its concerned functional areas. In addition, physical assets cannot be effectively managed on a tactical or operational basis only. Physical assets really have to be managed from a strategic level and using a strategic approach as well; hence the need for a holistic philosophy based on an organization-wide physical asset management system.

"Any questions?" Darren asks after a few minutes in a serious tone.

Karen is the first to raise her hand. "Don't organizations already have a system in place?" she asks. "How have they been managing physical assets all these years then?"

"Great question, Karen," Darren replies. "Many organizations do have a physical asset management system in place, and staff is quite familiar with the concept. Some organizations have rudimentary physical asset management systems that are able to fulfill basic activities such as identifying, tracking, and maintaining physical assets but not necessarily in a formalized and structured manner."

He stops and looks around to ensure everybody is with him. He then continues: "This approach to physical asset management is acceptable up to a point, especially for organizations operating in less complex and/or competitive environments. As organizations expand and become more asset-intensive, and as the environment becomes increasingly more competitive (in some cases regulated), the need for a formal, structured, and effective physical asset management system is critical."

"OK," Nihom suddenly interjects. "If I understand correctly, organizations need to have a physical asset management system for better direction and control in coordinating the many facets of managing physical assets in the constantly changing business landscape. Is that a correct statement?"

"Well, if you look at the ISO 55000 standards, you'll see that an asset management system is defined as a set of the interrelated and interacting elements of an organization, whose function is to establish the asset management policy, the asset management objectives, and the processes needed to achieve those objectives," Darren replies right away.

He then walks toward the whiteboard and starts drawing while talking: "In a nutshell, the physical asset management system is the link between the management of the organization and the organization's physical assets to realize value. This relationship is shown in this diagram" (Figure 10.1).

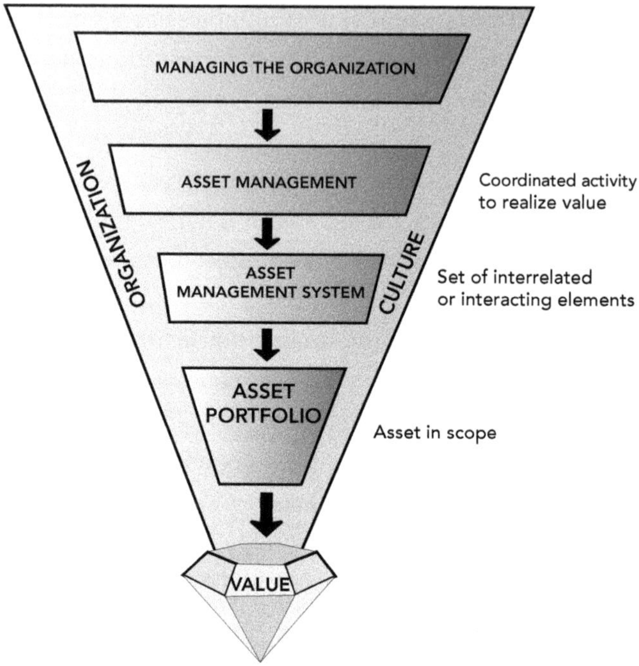

FIGURE 10.1 Organization and physical asset management relationship.

While everyone is contemplating the diagram, Darren picks another spot on the whiteboard and lists the following:

- Context of the Organization
- Leadership Commitment
- Multi-level Planning
- Support
- Operational Control
- Performance Evaluation
- Continual Improvement

Pointing at the list, Darren explains: "ISO 55000 further identifies seven sets of requirements for an asset management system in an organization. Meeting those requirements provides direction, alignment, coordination, collaboration, control, and continual improvement."

"Should an organization meet all these requirements?" Jerry asks inquisitively.

Darren looks at him and smiles. Then he says: "Actually, ISO 55002 states that compliance with all the requirements of ISO 55001 represents achieving the minimum standard for an effective asset management system and should not be seen as the final goal. In many ways, most organizations these days do not have any control over, nor see the need for, compliance with ISO 55000. However, this could change at any time in the different industries of the public and private sectors. Successful organizations grow and evolve naturally, and as the internal and external business environments change, so organizations have no choice but to adapt to the situation. They adopt new technologies and approaches to reflect the higher demands of the environment and meet new requirements. As business environments and demands are quite frequently subject to change, there is a clear need to adopt a formal and structured approach to physical asset management."

"I have seen many organizations which have good structured systems to manage their physical assets," Zach says vehemently. "Is there anything they are missing or not doing well?"

"True," Darren replies. "However, the reality here is that even with formalized and structured asset management systems in place, organizations still face different challenges impacting functional areas in different ways and ultimately affecting the realization of value."

He pauses for a moment and looks at the ceiling. "It follows, therefore, that an effective physical asset management system has to be designed and operationalized carefully with due analysis and evaluation of a range of different strategic/tactical/operational considerations."

He turns to the audience and adds while waving the handout over his head: "Take a few minutes now to read the second page of your handout."

Second page of the handout:

- Physical asset management operates at different levels in the organization. Strategic planners consider strategic physical asset-related decisions, managers deal with the tactical activities, while operational staff takes care of the operational tasks.
- The strategic/tactical/operational levels in organizations do not function in isolation. A physical asset-related decision or activity at one level

can impact directly other decisions/activities at one or more other levels. Individual decisions/activities should not be considered in isolation.

- Physical asset management systems require a governance framework to be sustainable and to reflect the ongoing interrelated relationships, factors, and other influences upon the system. The governance structure rearranges and delineates power and governing roles and responsibilities within the organization.
- Most, if not all, organizations cannot operate without physical assets. Physical assets are all around and are used in organizations at all levels and across all functional areas. Furthermore, individual physical assets cannot be considered in isolation, as they are part of the physical asset systems/subsystems to realize value. The physical asset management system therefore must adopt the broader view of systems/subsystems.
- Establishing a comprehensive physical asset management system can be a major undertaking for an organization. The asset management function, if in place, must be backed up with enough funds and resources to be able to operate efficiently. This can become quite expensive.
- Physical asset management systems have to be implemented across the whole organization. Physical asset management activities happen across all levels of the organization, across all functional areas within the organization, and across all physical asset types.
- Physical asset management systems can only be successful if they are consistently implemented across the organization and staff uses the policies, procedures, and approaches at all times rather than finding ways of overriding them.
- Physical asset management systems are only as reliable as the range of analytical tools and techniques used to analyze different physical asset-related scenarios and the implementation of the decisions/results derived from them.
- Physical asset management systems ideally should be regularly assessed or audited to ensure they are being operated in compliance with organization policies and procedures or any other regulation, if applicable.

Once everybody seems to have read the material, Darren continues, "These are some typical factors to consider when exploring the idea of an effective physical asset management system."

"This very much makes sense," Karen replies in a delighted voice. "It is useful information that I can include in my presentations."

Darren gives her the thumbs up.

CHAPTER 11

CHARACTERISTICS OF AN EFFECTIVE PHYSICAL ASSET MANAGEMENT SYSTEM

Darren takes a sip from his water bottle and addresses the audience. He asks, "Are you ready to shift gears and get into some more interesting stuff?"

Everyone seems to be in agreement, as there is no response. And so Darren continues: "By now we know what the requirements for an effective physical asset management system in an organization are as recommended in the ISO 55000 standards. We also know some of the key factors to be considered when developing and implementing an effective physical asset management system."

Nihom stands up and says, "In my organization, management is supportive of developing and implementing a physical asset management system. However, I need to know what I should be doing next."

"Good to hear that you are getting the support to move ahead. Many are not as lucky as you are, because this is a major step forward, getting the buy-in from senior management. So the next step for you is to consider the characteristics of an effective physical asset management system."

Darren pauses before putting out this question in a loud voice: "What features should an effective physical asset management system display?" And without waiting for any answer, he resumes.

"Let me give you some examples," he says while proceeding to the whiteboard where he writes down:

1. ***It must be reliable and shown to be effective within the constraints set by the organization.***

   Darren turns around and continues: "The physical asset management system does not have to be perfect and be without flaws—and let me tell you, it will never be!—but it does have to offer the level of coordination and alignment for which it was designed to meet the minimum acceptable requirements."

   Darren waits for a few moments in case anybody has any questions. Then he proceeds with writing the next feature on the board.

2. ***It should have an organization-wide scope.***

   "In most cases, the physical asset management system should be designed and integrated with the whole organization in mind," Darren explains. "But it must also include external environmental factors that may come into play such as industry regulations, market conditions, competitors, stakeholders, and even natural phenomena like climate change."

   He turns around and writes the third feature.

3. ***It must be practical and simple to understand.***

   "The asset management system has to be practical and implementable," he adds. "This is important, because if it doesn't meet these two basic criteria, people will not buy into it or will only be engaged to a limited extent."

   Someone at the back of the auditorium raises her hand. "Can you please elaborate on that?" a thin feminine voice asks. "It looks like a very important element."

   Darren barely hears the full question, but he understands what was asked. "Absolutely," he swiftly replies. "It is indeed an important feature of an effective physical asset management system. In many organizations it's common to encounter physical asset assessment systems with processes and procedures that are so complex and confusing that people don't embrace them or don't know how to use them properly. Try to keep them simple and practical."

   "The next feature of an effective physical asset management system is that it must be realistic," Darren adds, and he writes it on the whiteboard.

4. ***It must be realistic.***

   "The physical asset management system should not treat all assets in an organization the same way—that would just be a waste of time and effort. Operating context and consequences of failure for different asset classes must be taken into consideration."

   Darren scans the auditorium for any reaction and, in an authoritative manner, concludes: "The physical asset management system must be selective and targeted to the most critical assets based on their different operating context."

   Again no reaction, as everyone is busy understanding and absorbing the vast amount of information. He goes on to write the next feature.

5. ***It should meet the requirements of internal and external standards or applicable legislation.***

   After writing the fifth feature, Darren proceeds to explain: "Ideally the physical asset management system should be in compliance with existing internal and external procedures, standards, and legislations. Depending on their operating context, many organizations may choose to develop their physical asset management systems to meet the minimum requirements, whereas others may go above and beyond the minimum levels set by the applicable standards and regulations."

   Darren writes the next feature on the board and faces the audience.

6. ***It must add value.***

   "Comprehensive physical asset management systems can be expensive to develop, implement, and sustain," he starts to explain. "The system should balance out the weight of the high cost by adding value to the organization. It can be challenging to measure the value-added aspect or justify the cost-effectiveness of a physical asset management system, because its entire output is largely unnoticed; there are no tangible results produced. However, its contribution to helping organizations make better decisions and deliver on its strategic objectives through alignment and coordination has to be recognized."

   Darren looks around the auditorium. Most participants have their heads down taking notes. He likes it, as he sees the interest in the information being provided.

He slowly starts writing down the seventh feature.

7. ***It must have a full lifecycle approach.***

   He waits for a few moments until everyone is paying attention to the whiteboard and clarifies: "The physical asset management system must be designed and applied across the whole lifecycle of physical assets and for each different class as necessary. This could be the key linkage to the strategies and activities across the different functional areas and the whole organization."

   "The last one is straightforward!" he exclaims while writing on the board.

8. ***It should be rolled out as a program.***

   "The design, development, and implementation of a physical asset management system must be managed as a program containing a set of related projects. The program will require support and resources from the organization to be able to function as effectively as possible."

"A program?" Zach quickly asks. "It looks like we need to hire an experienced program manager, or maybe I should get some more project management training to take this task on."

Darren smiles at Zach, most probably because he admires his enthusiasm, and responds, "I cannot comment about hiring, but I strongly recommend that sound project management skills are essential to successfully developing and implementing an organization-wide physical asset management system."

"Can you please elaborate a little more on that?" Jerry, who has been quiet for a while, asks. "I may include some of your ideas in my work program."

"Let's face it," Darren says as he glances at the audience. Then he starts to pace as he stares at the ceiling.

The participants brace themselves knowing that Darren is going for a relatively long explanation. He has done that a few times today.

Darren begins: "The physical asset management system that you would want to implement will have to be developed to work organization-wide and has to be implemented and operated in the same manner throughout the entire organization in order for it to function effectively. This means that the whole organization will be impacted and functional areas will be affected, some more than others. For the program to be successful, you will need to get a commitment to organization-wide physical asset management, and all stakeholders need to be engaged and fully bought in to the program and the physical asset management system idea. You may want to formally prepare and obtain sign-off for a program charter where you would scope the program; identify the program sponsor, champion, and major stakeholders; and outline the benefits and risks to the organization."

Darren turns around to pace in the other direction and continues with his explanation, saying: "Furthermore, from development to implementation, the whole program of the physical asset management system should be carefully scoped and planned. In most cases this will include a roadmap to identify all the related projects and critical milestones; a Gantt chart with all projects sequenced and scheduled; and cost estimates for each project. The planning process should also consider the assignment and availability of resources for the development and implementation of the physical asset management system."

He stops in the middle of the auditorium and continues while clasping his hands. "The program," he explains to his listeners, "will most probably be one of the many initiatives the organization is working on, and you will want to make sure adequate staff time and resources are dedicated to the program. The last thing you want is staff burnout and loss of interest and motivation in the different projects of the program they are involved in."

"That sounds interesting so far," Jerry interjects. "I think that with some additional training, I should be able to manage such a task."

"Yes, Jerry," Darren agrees. "It is a very interesting undertaking in general. I enjoyed my time when I led such a program in my past job. But the real 'fun' is during the actual implementation, where you will have to 'walk the talk' to make things happen. You will have to show progress and results, and ensure that all aspects of the design and implementation of the system are carried out effectively while keeping the team engaged and motivated. There will be plenty of challenges along the way, and you need to keep the momentum and the integrity of the whole program going. This is one of the main reasons why the design and implementation of the physical asset management system should be allocated as a specific responsibility to a suitably qualified program lead, or even an external specialist to provide expertise, counseling, and, more importantly, *thought leadership* on physical asset management."

Darren stops for a few seconds to catch his breath and have another sip of water.

"That's not all!" he quickly cautions. "Once execution is complete or approaching an advanced stage, ownership becomes an issue. Control and management plans need to be in place with designated responsibilities to ensure the program is sustained and meeting the set objectives. However, as the adage goes, 'everyone is an asset manager'; each and every manager or staff member from different functional areas in the organization must take responsibility for managing the physical assets in his or her area as prescribed in the physical asset management system."

He looks around and senses everybody is still following, and so he continues: "As with all projects leading to business transformations, the longest timeline is to change the organizational culture. To be truly effective, the organization's physical asset management system must be embedded into the commonly held beliefs and values that shape the organization's culture. Continual staff training and development are necessary in the post-implementation stage. There should be processes in place for program evaluation and recalibration of the physical asset management system on a regular basis."

Darren slows down and looks at Jerry as he says, "The project lead must keep tabs on the change management aspect of the program and should be given the necessary authority to ensure that this process can take place."

Jerry feels the pressure of Darren's look. "I am really excited," Jerry declares with an enthusiastic voice. "The way you have explained the aspects of the program and elements of the physical asset management system does help a lot. I feel more confident now to take on this challenging task."

"Glad to hear that you are feeling very confident, Jerry," Darren responds. "We can meet later to discuss a few other aspects of the program that you need to think about before getting right into it."

Then he turns to the class, takes a peek at his watch, and says: "It's now time for us to bring together everything we have discussed so far with regard to strategy execution, organizations, functional silos, systems, and the dynamics of existing interrelations, interactions, and interdependencies to realize value from physical assets."

"You remember the diagram I showed you when we discussed interrelations, interactions, and interdependencies?" Darren asks while he searches for a file on his laptop.

After a few seconds he pulls up a diagram on the screen and continues: "This is a slightly different version of the diagram I previously showed you. During implementation of an organization-wide physical asset management system, interrelation will be the *physical asset lifecycle phases*, and interaction will be the different *functional areas*" (Figure 11.1).

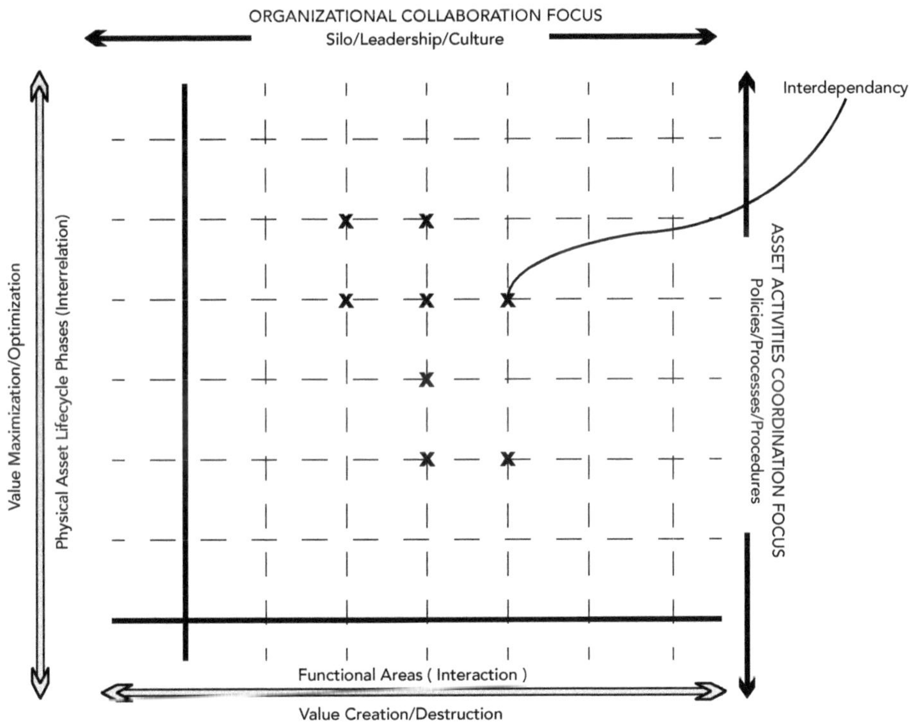

FIGURE 11.1 The organization-wide physical asset management system.

Darren then points his laser at the horizontal and vertical axes and continues: "When we apply the above approach to the concept of quality, flow, and involvement from the book *Quality, Involvement, Flow—The Systemic Organization*, we have the following." And he flips to another slide with three definitions:

- INTERRELATION is the flow from different physical asset lifecycle phases—flow of information via policies, procedures, and practices.
- INTERACTION is the involvement of all individuals, functional areas, and departments to collaborate and share information.
- INTERDEPENDENCY is the quality of interrelation and interaction to achieve the goals.

"It's this complex network of interdependencies," Darren explains, "that is critical to tackle during implementation of an organization-wide physical asset management system to effectively coordinate the act of flow, involvement, and quality across the organization."

He takes a pause, looks again at his watch, and says, "It's time for a coffee break. You all deserve it! Let's reconvene in fifteen minutes to wrap things up."

CHAPTER 12

MOVING TOWARD ORGANIZATION-WIDE PHYSICAL ASSET MANAGEMENT

Darren makes his own coffee in his office and knows exactly how strong he wants it. He sits at his desk and starts checking emails while sipping his coffee. He checks his phone—no missed calls and no messages. He quickly types a text message to Debbie to let her know what time he will be home tonight. Darren dedicates his Thursdays to cycle training. He usually gets home at around eight o'clock. So it will be a long day for Darren, but he's used to it. He looks at the time to make sure he's not late for the seminar restart. He grabs his coffee mug and heads to the auditorium.

"By now we know that organizations need an effective physical asset management system to manage their physical assets and to meet their objectives in the context of their operations. We very often hear that organizations have been managing physical assets well for a long time. If you ask any one individual in an organization, you will hear that things are indeed working well. However, if you bring all those individuals from that organization together in one room and ask the same question again, you will probably get a completely different answer." This is Darren's opening statement after the coffee break. It's a powerful statement with the intention to challenge the members of the audience and make them think.

"Before anyone asks me 'why,' let me tell you," he says in a firm voice, "that the reason is that physical asset management systems tend to be *silo-based,* in that they have been developed by a specific department or functional area for specific use in that particular department or functional area. For example, an organization may have developed a highly effective financial asset management system that only operates within the financial department. Now, this same financial asset management system may have little to no use or effectiveness within the maintenance management function. This idea of an isolated, function-specific asset management approach is the basis of the traditional functional silo syndrome. Nowadays when you look around in organizations, you can see many of those types of function-specific asset management systems popping up, such as IT asset management or fleet asset management systems."

Zach raises his hand. "This reminds me of the definition of mechanistic organization which states that the results of the whole organization equal the sum of its individual, separate, and hierarchical parts," he says in a calm voice. "And you did mention that we should move away from that to be more effective."

"That's correct," Darren confirms. "The effects of the isolated, function-specific asset management systems (mechanistic approach) are widespread in many organizations and have significant implications. Either the functional areas cannot see the bigger picture, or they have been instructed to operate that way. The problem is that organizations are investing lots of money in building physical asset management systems, but not many of them are implementing it as required."

"What do you mean?" Zach asks, confused.

"Let me explain," Darren says. "Implementation of an organization-wide physical asset management system must be broken down into two distinct stages."

He goes to the whiteboard and writes "AM framework," then adds, "The first stage is to implement the asset management framework with a focus on the strategies, policies, and tools."

Darren then writes "Transform the organization" and explains, "The second stage is to transform the organization through a systems approach with a focus on the network of interdependencies to create value."

"Lots of documentation, training, and other resources are available pertaining to the policies, strategies, plans, tactics, and tools needed for a physical asset management program. Many organizations have most of these partially or fully developed and implemented."

Darren looks around the auditorium and continues in a stern tone: "However, let's face it—these do not transform the organization, per se, nor do they create value on their own. They are transformation enablers and tools that can be applied to drive change in organizations to create value."

"I can relate to that," says a gentleman sitting in the fifth row on the right-hand side. "Our organization spent hundreds of thousands of dollars to develop policies, strategies, plans, and what not—all sitting in binders—but we are still struggling to implement them. After almost two years, I personally have not seen much change." He sounds very frustrated.

"I fully understand where you are coming from, and there is much more to it," Darren replies in a reassuring voice. "What really creates value in physical asset management systems is the new perspective of systems thinking, which is that every organization exists to provide value as a whole, through coordination and collaboration across functional areas. Organizations need to undergo a paradigm shift focused more toward a systemic understanding to make that transformation process happen. In other words, *to move toward an organization-wide physical asset management system, the focus must be on the transformation process.*"

"Can you please explain a little bit more?" Zach asks, confused.

"First let's take a step back and look at the shortcomings of a silo-based approach in implementing physical asset management systems across organizations," Darren starts to explain. "With an organization-wide approach, organizations are able to effectively deal with physical asset-related events and decisions. This helps to achieve the strategic objectives and realize value from physical assets. On the other hand, with a silo-based approach, organizations manage a limited number of asset-related events and decisions within specific functional areas, without any recognition of the impact on the whole organization's objectives. This approach of physical asset management in silos is limited with significant shortcomings. It does not consider physical assets in a holistic sense or how they are linked to opportunities to maximize value."

"Now let's take a look at the shortcomings," Darren adds.

He picks up the marker and writes the word "gaps" on the board. "The first one is *gaps*," he says. "The silo approach creates potential for gaps. Physical asset management in any sense means managing physical assets over their whole lifecycle phases from planning and acquisition through to disposal in order to provide maximum value to the organization while balancing costs, performance, and risks. However, a fragmented approach creates the potential for gaps, and/or raises the possibility for assumptions to be made, thinking that specific physical asset management activities are being adequately handled by other departments in the organization."

"Can you provide an example?" Zach asks, somewhat confused.

"Sure," Darren says. "Let's consider the decision to cut back on the maintenance budget, which is quite a common theme these days in many organizations."

Darren pauses and looks at the audience. He sees many heads nodding and continues with a grin: "Although there may be valid reasons for such reduction, it must be considered not only in terms of the savings on paper but also in terms of the consequences if things go wrong. Organizations need to evaluate the consequences of failure of physical assets due to maintenance budget cuts. The impact on the organization's ability to continue to deliver value to its stakeholders must be assessed before validating the budget cut."

He then turns round and jots down "overlaps" on the board. He explains, "In contrast, the silo approach can also lead to *overlaps*, where obvious physical asset management activities are over-managed. If a physical asset or asset system is critical to operations, several functional areas will be very concerned about it and its overall impact. In that situation we may find a number of different functional areas carrying out similar activities or duplicating efforts around that same physical asset or asset system and sometimes in a contradictory manner. For example, the maintenance function may be spending a significant amount of money in fully refurbishing a critical piece of equipment while engineering in conjunction with finance may be planning to replace it in this year's budget."

Darren pulls up a diagram on the screen (Figure 12.1) and says, "This diagram, taken from *Physical Asset Management: An Organizational Challenge*, illustrates the potential gaps and overlaps between different functional areas in an organization over the whole lifecycle phases of physical assets."

Darren takes a lengthy pause to give everyone enough time to decipher the diagram. In the meantime, he takes a couple of sips of his coffee.

After a few minutes, he continues: "The second area where the traditional approach to physical asset management fails is when an event develops that considers the physical asset on its own rather than considering a number of different factors in the broader context of the organization's objectives. Business decisions

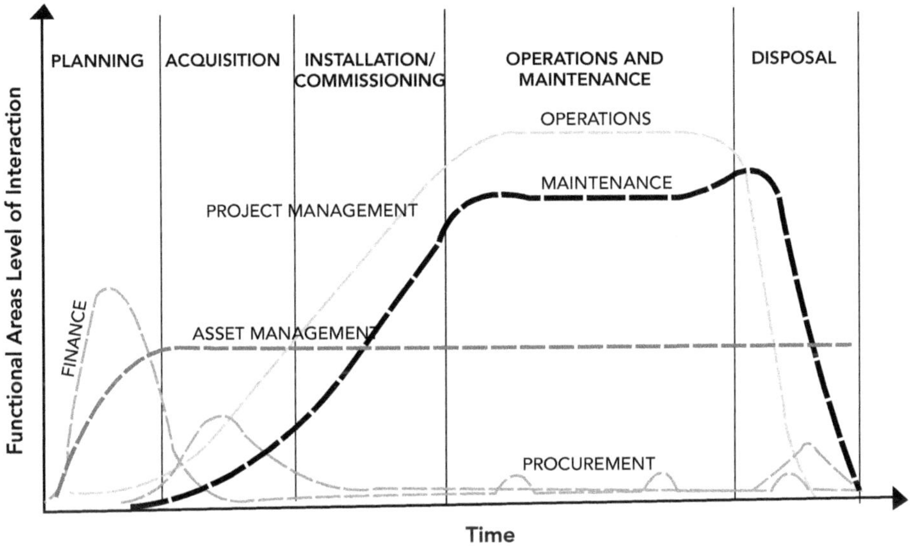

FIGURE 12.1 Gaps and overlaps.

can impact physical assets through an array of different functions, and they can also impact each other in different physical asset systems. In most applications, physical assets are interdependent and do not simply operate on their own without affecting one another. A failure to understand the interdependent nature of physical assets leads to failure in understanding physical asset systems and their operational context. This, in turn, leads to an unreliable and ineffective physical asset management system that is unlikely to help an organization in accomplishing its organizational objectives."

Darren takes a brief pause in anticipation of questions. Hearing none, he adds, "To illustrate this concept let us consider a truck on a mining site."

He pulls up another slide (Figure 12.2) on the screen and comments: "The truck on its own sitting in the garage does not bring any value (for operations) to the organization even though it costs money (value for finance). When we look at the same truck in the loading and hauling process, we can see how the truck fits in and is important to making that system perform its function (value for operations). Making any change to the truck performance can affect the performance of the loading and hauling system."

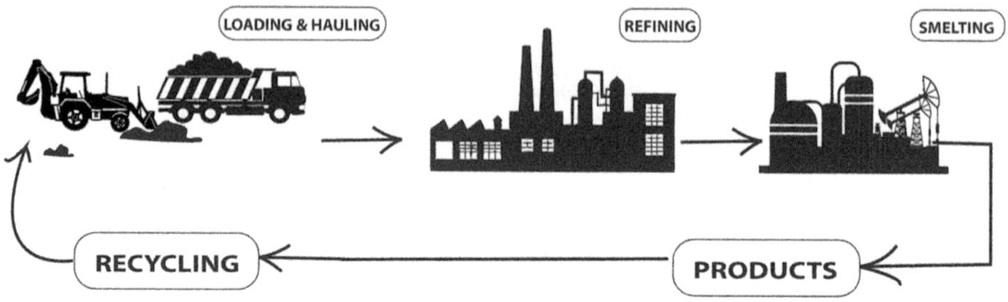

FIGURE 12.2 Assets and asset systems.

Darren moves his laser pointer over the slide and at the same time explains: "Now if we picture the truck in the loading and hauling system within the context of the whole mining site, then we can better appreciate the overall operational context (value for organization). Any change to the truck's performance will affect the loading and haulage performance, which, in turn, will impact the other upstream and downstream processes."

Darren puts down his laser pointer and takes a few steps toward the audience, adding: "Finally, the third way in which silo-based physical asset management can negatively impact an organization's performance is through lack of understanding. Nowhere is this more apparent than in physical asset valuation, where different functional areas perceive the value of physical assets differently. Finance will value physical assets based on accounting principles, and a different valuation will be used by risk and compliance staff for insurance purposes, whereas the maintenance function will value the physical assets from a performance and condition perspective."

He pauses and stares at Karen and then continues: "Worse is the case where operations, while knowing the limited capabilities of physical assets, demands that they perform consistently without willing to invest in them. Such an approach is unsustainable. One of the benefits of organizational physical asset management is in aligning these perceptions and views in pursuit of achieving the organizational objectives."

Darren turns around and starts pacing, saying as he paces, "However, before the organization can start to consider the gaps, overlaps, and their impacts, it's befitting to start defining both what a physical asset is and what it means to the organization and to develop a common understanding so that physical assets, their terminologies, and the value they deliver are clear.

"But everyone in an organization knows what a physical asset is!" Zach says, irritated.

"I'm not so sure about that," Darren replies. "The definition of physical assets for a maintenance manager can be significantly different from that of financial staff as is reported in the organization's financial statement. Have you ever sat in a team meeting where your engineering, financial, maintenance, and operational colleagues are all speaking a different language? They are all aware of their responsibilities and know the importance of their contributions in the decision-making process, but to align their viewpoints, terminologies, and objectives has always been a struggle."

Darren goes back to his laptop, pulls up another slide on the screen, and continues: "There are many processes and activities taking place around physical assets in different functional areas across the organization that make it difficult to understand who does what, when, and why. This creates a major hurdle and ultimately impedes the decision-making process."

"This is what I call the 'asset management spaghetti diagram,'" he says, pointing at the slide (Figure 12.3).

Everyone seems confused by the diagram. Not only is it hard to read, but it's also difficult to understand, hence the name "spaghetti." On the other hand, Darren seems excited at the looks on people's faces.

To put an end to the confusion, Darren starts explaining: "The diagram displays many of the activities related to physical asset management that are happening in organizations. These activities contain sets of processes, 'owned' by different functional areas, and performed by staff to meet the organizational requirements. However, there are challenges in harmonizing those activities/processes and aligning the functional areas to work together as an integrated establishment."

"Why is that so?" Karen asks.

"The reasons are many: poor communication channels between the functional areas resulting in poor information sharing, lack of transparency, and lack of internal coherence, among others," Darren replies.

"So how do we deal with that?" Karen asks in response.

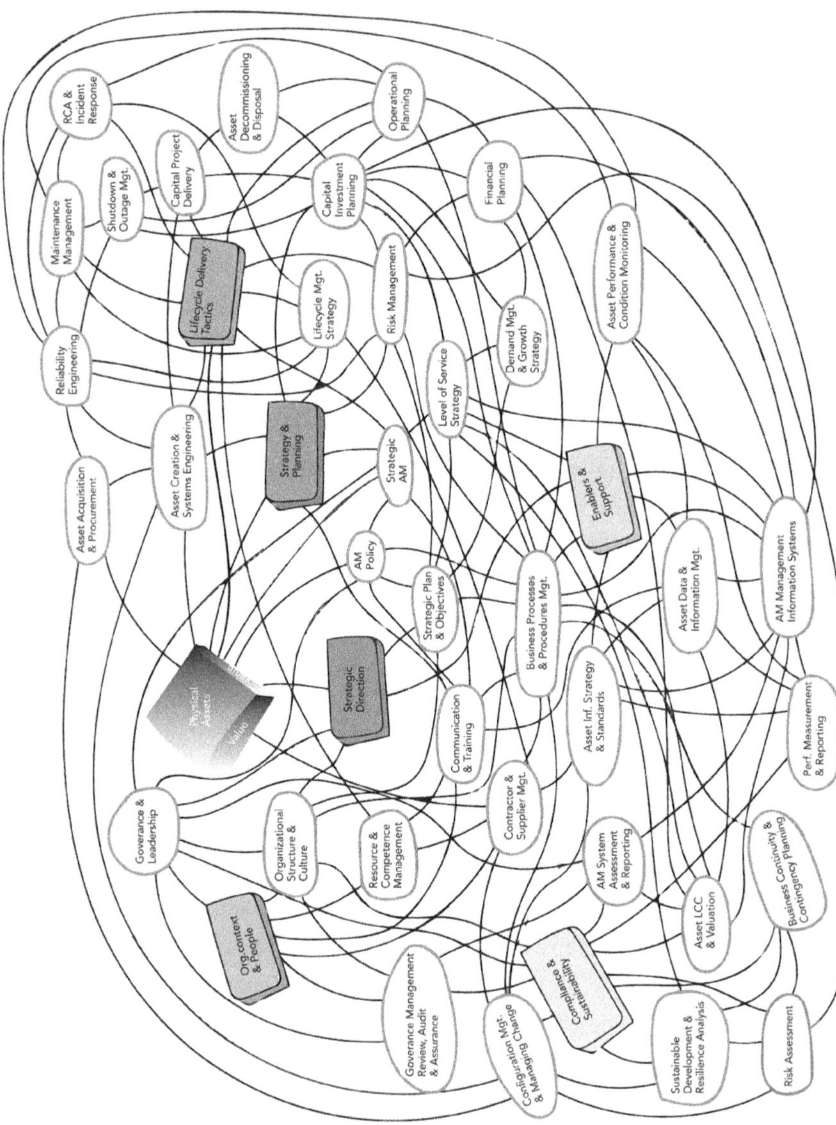

FIGURE 12.3 Physical asset management spaghetti diagram.

"Well, if you remember, earlier we talked about three areas where the approach of physical asset management in silos has significant shortcomings. We also said that we need to have a better 'common understanding' of physical assets and their activities before we can analyze gaps and overlaps in organization-wide physical asset management systems."

Everyone seems to agree with what Darren just said. He continues, "The 'common understanding' is intended to forge better communication, synergy, and collaboration among functional areas dealing with a wide range of physical asset classes across the organization."

Karen raises her hand and says: "I recently read an article about silos in organizations, and it does talk about assessing the degree to which an organization is siloed by measuring its *modularity*. If I recall correctly, it defines *modularity* as the ratio of communication within a functional area to communication outside the functional area. If the ratio of internal to external communication exceeds a certain number, then the functional area is determined to be siloed."

Darren nods his head in approval and adds with a smile: "That's very interesting, Karen. Thank you for sharing."

He then uses the clicker to switch to another slide (Figure 12.4).

"This diagram," he starts, "represents the management of physical assets in the context of organization-wide physical asset management (OWPAM) with linkages of the traditional functional areas to ensure physical assets realize value on an ongoing basis."

"This more holistic and macroscopic perspective removes physical assets out of the context of the functional areas' control and elevates them to the context of the whole organization. It also considers physical assets in systems and how they integrate with the organization's operational context to effectively create value rather than in isolation."

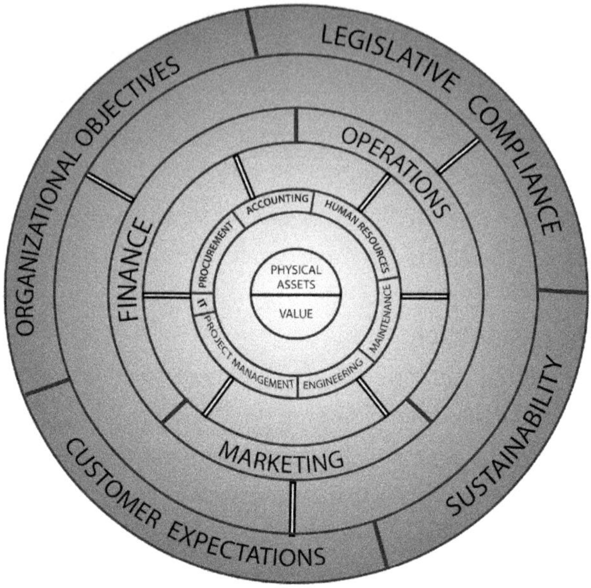

FIGURE 12.4 Asset management in the context of OWPAM.

Darren scans the audience to check if everybody is following. He then adds: "This obviously highlights the importance of physical assets, which makes them a key factor for consideration in strategic business decision processes. All of this subsequently develops an awareness for physical asset lifecycle phases and their interdependencies with functional areas across the organization. This in return raises the possibility of developing a full organization-wide physical asset management system. Therefore, organizations can improve their physical asset management system by looking across the functional areas and physical asset lifecycle phases to identify and assess the gaps and overlaps, as well as their impacts, using a consistent approach."

"Wow!" exclaims Jerry. "That is an interesting statement, especially the idea of organization-wide physical asset management."

Darren looks at him and smiles. He does not say anything, but deep inside he is glad that the message is getting through.

"How do we make the transition from where organizations currently are to an organization-wide physical asset management approach?" Jerry asks.

"Good question, Jerry. I am glad you ask, as this is the core theme of this seminar."

Darren takes a sip of coffee, fixes his tucked shirt, and walks toward the front row. He then suggests, "Let's see what it takes to move toward an organization-wide physical asset management approach."

Everybody in the auditorium goes totally silent again and waits for Darren to continue.

"Different organizations," Darren notes, "have a multitude of different approaches to physical asset management, ranging from the traditional silo-based physical asset management system to a fully integrated organization-wide physical asset management system. Organizations should ideally aim for an organization-wide physical asset management philosophy that supports a physical asset management system designed and implemented with the organization's strategic objectives in mind, and as a main priority. To move from one end of the spectrum to the other is a long and tedious transformation process. First of all, organizations need to know where they are in the spectrum. Then they need to be patient and take a well-staged modus operandi with a focus on specific stages of the process as suited to their needs to move forward."

Darren switches to another slide (Figure 12.5) on the screen and continues: "This diagram gives you an idea of the typical stages involved for organizations to move from the awareness stage of an organization-wide physical asset management philosophy to excellence. It is a tool commonly used to assess the maturity of an organization's physical asset management system."

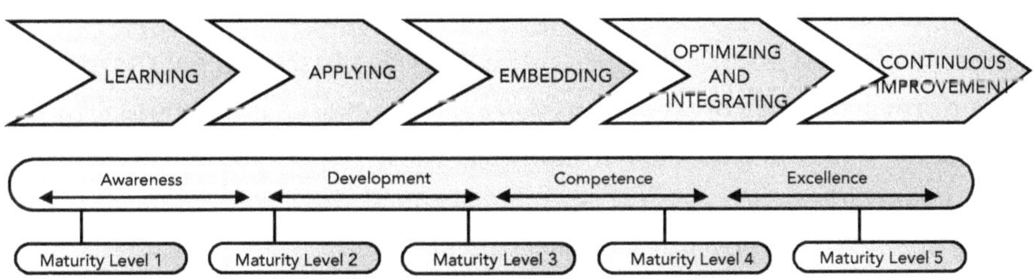

FIGURE 12.5 Asset management maturity scale.

"As I said, to move from one end of the spectrum to the other requires a transformation process."

Darren moves his laser pointer from the left-hand side to the right-hand side of the screen. He then flips to the next slide (Figure 12.6) and continues, "The transformation process involves moving away from the silo approach, improving communication, assessing physical assets across the organization, and, ultimately, aligning the functional areas to collaboratively manage those physical assets."

Once more, Darren runs his laser point across the screen from left to right to demonstrate what he is trying to convey.

He starts at the left of the top row and says, "These are the stages of the transformation process to develop an organization-wide physical asset managemen system," he says.

He then points to the bottom row and says, "These are the organizational characteristics that will have to change to enable the transformation process."

Darren gives the audience a few minutes to analyze the diagram and assimilate what he just said. He takes the opportunity to check his phone and have a sip of coffee, and then he resumes, explaining, "Toward the end of the process, physical asset management becomes a corporate initiative, rather than buried in one of the functional areas, and forms an integral part of the strategic planning process to help develop the organization's strategic and asset management objectives."

"Organization-wide physical asset management is about making the organization better at managing physical assets over their whole lifecycle through the collective and coordinated capabilities of individuals, functions, and operations across the organization. It improves the organization's likelihood of performing at the level necessary to achieve its strategic objectives. Organizations that are able to manage physical assets over their entire lifecycle phases in a *holistic* manner will perform better than organizations that are not. Some considerations when an organization is deciding how much effort to invest in developing and implementing an organization-wide physical asset management system are complexity, size, and the organization's physical asset investment."

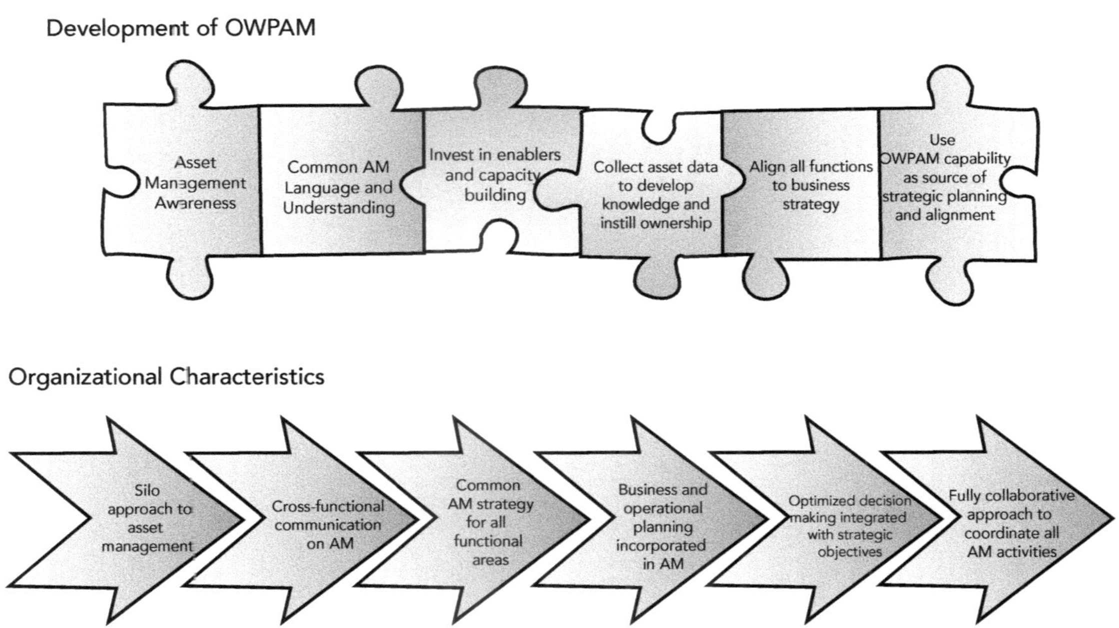

FIGURE 12.6 Development stages of OWPAM philosophy.

Darren takes a deep breath and continues: "Organizations which are physical-asset-intensive with highly critical assets may invest considerable sums in physical asset management and be prepared to invest more in order to develop a fully comprehensive organization-wide physical asset management system. This is because the consequences of not doing so may be significant. Similarly, organizations which are mandated to deliver essential services need to have a fully operational organization-wide physical asset management system in place to manage their infrastructure assets in a more *holistic* manner."

"Sorry to interrupt," Nihom says while waving his raised hand. "It's the second time you have mentioned the word 'holistic.' Is there anything in particular you are trying to convey?"

Darren smiles and nods his head in agreement. "Good point. Actually yes, I would like to elaborate a little bit on what I mean by 'holistic.'"

"By definition in the *Oxford Dictionary*, 'holistic' means 'characterized by comprehension of the parts of something as intimately interconnected and explicable only by reference to the whole.' In philosophy, the holistic principle emphasizes priority of a whole over its parts."

"Hold on a second!" Zach almost shouts. He quickly flips through the pages of his notebook and stops. "That's almost identical to what we came up with in our definition of organization and system!" he exclaims in a cheerful voice.

Darren is amazed at what he is hearing. "I have to say that I am very impressed with the questions that are being asked and the way all of you are processing this vast amount of information. I can see that the future of physical asset management will be in very good hands."

"Yes, you are correct," Darren continues as he glances at Nihom. "What I am trying to convey here is that when we talk about the discipline of 'holistic physical asset management,' we traditionally mean managing physical assets over their entire lifecycle—'from cradle to grave,' as it is more commonly described. However, that argument leads to the question of whether the full potential of physical asset management is achieved by focusing solely on the physical asset lifecycle."

Darren clasps his hands in front of him as if he is starting to pray as he asks: "Is the 'holistic' approach only applicable to the physical asset lifecycle phases? What about the processes and the people who manage the processes? Shouldn't they have a 'holistic' approach as well?"

It is the same deafening silence again, and Darren thinks to himself that this might be a little bit too much for everyone after such a long day.

Nevertheless, he continues, knowing that he will soon be wrapping it up for the day. And so he says: "As we saw earlier, physical assets in organizations are managed throughout their whole lifecycle by many different functions dealing with different portions of the lifecycle: operations operates, maintenance maintains, project management does the design, etc. In the organization-wide physical asset management context, the objectives are to address the interconnections and alignment of these functions by getting rid of functional silos and instill a veritable 'holistic' approach."

Karen puts up her hand. "Do you mean there is a difference between *holistic physical asset management* and *organization-wide physical asset management*?" she asks, perplexed.

"I would not dwell too much on the difference per se," Darren replies promptly. "What I am trying to say is that organizations historically have been applying *holistic physical asset management* as the holistic approach to managing physical assets with a focus on managing physical asset lifecycle phases and their activities. In my opinion, the real holistic approach to managing physical assets should focus on the functional areas' collaboration as well, hence an organization-wide physical asset management approach."

He stops to see if Karen wants to add anything. Seeing no reaction, he continues: "Let me explain further. The traditional approach of physical asset management typically involves discrete functions—namely, finance, operations, and maintenance. Organizing in such a manner creates the potential for functional silo syndrome resulting in omissions and duplications or gaps and overlaps, as highlighted earlier. This fails to consider physical asset management objectives in their widest sense and from a strategic perspective. After thorough review of the challenges regarding strategic execution in organizations and the intricacies of systems thinking, implementing a physical asset management system is a much more complex endeavor."

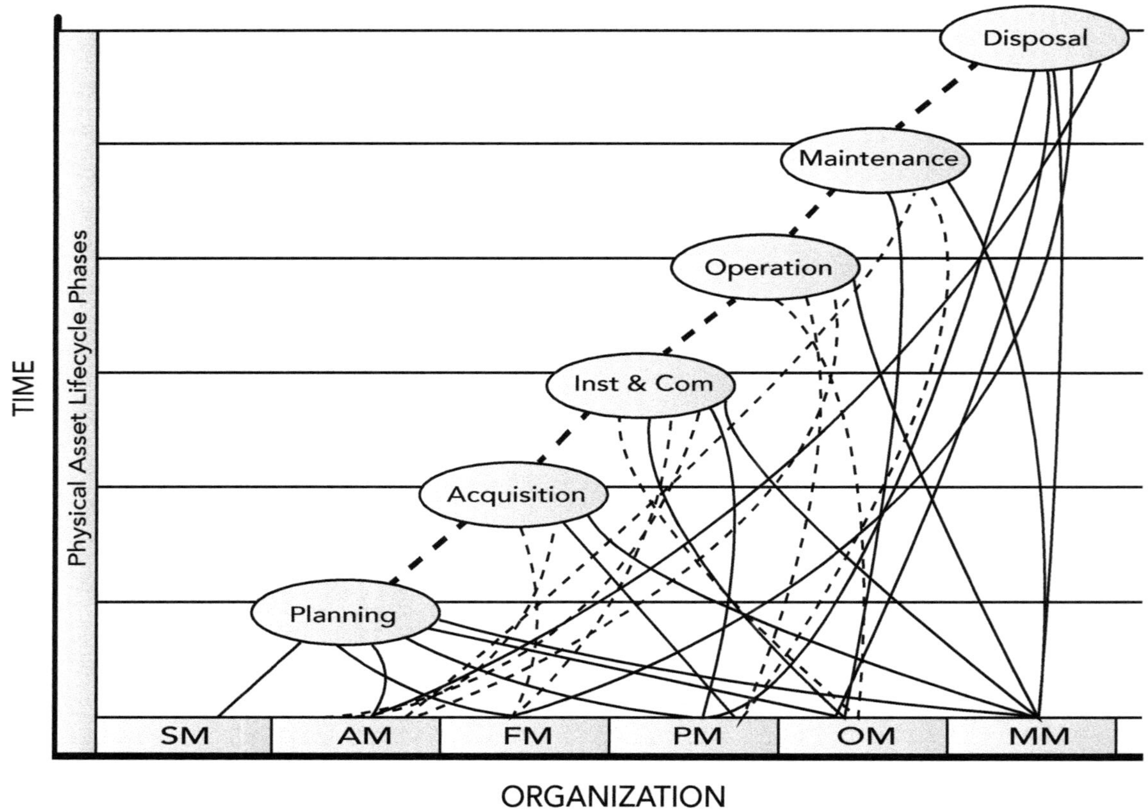

FIGURE 12.7 OWPAM chaotic situation.

Darren walks back to his laptop and scrolls to another slide, which he pulls up on the screen (Figure 12.7). "This diagram," he says, "illustrates the complex situation that exists. On one side you have the physical asset lifecycle phases' issues and requirements and on the other side the conflicting priorities of the different functional areas. Between them, there are so many activities happening at all times creating lots of interactions. Remember the 'spaghetti diagram' slide? This is what results when we put everything in perspective with respect to time and the organization."

He pauses to check if everyone remembers and then says: "Those activities and interrelations are key elements of a physical asset management system, but unfortunately they are too often dispersed and unstructured within an organization. Depending on the nature of the organization and the complexity of its physical assets, this can sometimes create quite a chaotic situation. To that effect, ISO 55000 clearly states that an asset management system and its related activities are an integral part of managing an organization, and they must all be aligned to achieve its objectives."

Raising his voice and pointing his right index finger up, Darren says defiantly, "No wonder why we should not look at physical asset management through the lenses of individual functional silos, but in an organization-wide manner!"

"This makes total sense to me," Jerry says while nodding his head. "Is it correct then to say that to successfully fulfill the potential of *holistic physical asset management* will entail aligning the activities of all functional areas over the entire life of physical assets to achieve common organizational goals?"

"Yes, you are spot on, once again," Darren replies. "For this reason, organizations need to shift toward an organization-wide physical asset management philosophy."

Darren looks at his watch and notices that he has just over five minutes left to wrap up the session. He does not want to hold anyone back. Deep inside he knows he has conveyed the core message of this seminar, which is the need to move to an organization-wide physical asset management approach. He also knows that he has provided a vast amount of information that will be significantly helpful.

To wrap up, he pulls up another slide with a few website links and names of documents.

"It's time now to wrap up the seminar. I see that you are grasping the organization-wide physical asset management concept really well. Here are some links and documents that you may find useful if you want to dig a little deeper into the subject."

People start to either take notes or use their cameras to capture a screenshot.

"Next time we meet, we will get into the details of organization-wide physical asset management from a system perspective. In the meantime if you have any questions, please feel free to contact me via email," Darren continues.

"Thank you all for coming, and have a safe journey home. Take care."

As the auditorium empties, he starts to pack his bag while thinking about his upcoming workout at the gym tonight.

CHAPTER 13

ORGANIZATION-WIDE PHYSICAL ASSET MANAGEMENT

Darren has planned his monthly block triathlon training for this weekend. In other words, he will do all three disciplines simulating a race. This is in preparation for his next race, which is a couple of months away, and he wants to make sure he is at his peak fitness level by then. That is his objective, and he has put together a strategy to achieve this goal. There is no backing off or taking it easy at any time. Along the way, Darren is very meticulous in monitoring his performance at each training session to ensure he continues improving his fitness level.

This Sunday morning after his usual 40 laps swimming in the pool, Darren heads out on his bike for a 30-mile ride at an easy pace to save some energy for the run. At this time of the day, there are few people on the road, and Darren makes the most of the ride. After the bike ride, he switches gears and takes on a 6-mile run—again, at an easy and comfortable pace. Darren always enjoys his training routine and feels good in his body once it is over.

After a copious and well-deserved brunch, Darren completes some of his routine household chores. He then sits in his den and pulls out the draft academic paper he started writing a few months back, which he has titled "Organization-Wide Physical Asset Management (OWPAM)." He regularly updates the paper based on new ideas and findings. This is part of his preparation in his quest for a doctorate. He knows he's got a long way to go, but he will not give up for any reason.

Darren makes himself comfortable with a bottle of water within reach and starts reading.

Organization-wide physical asset management offers a comprehensive approach to whole lifecycle physical asset management by aligning people, processes, and activities from the different functional areas to achieve the organizational common objectives with the help of technologies.

Organization-wide physical asset management is a multi-dimensional system approach with a focus on the vast array of interdependencies between activities happening in physical asset lifecycle phases and in the different functional areas. For an organization-wide physical asset management system to be successfully implemented, we need to consider all the functional areas dealing closely with physical assets and their related activities across the whole lifecycle phases of physical assets.

Physical Asset Lifecycle Phases

There are a lot of activities that need to happen over the entire lifecycle of physical assets to reap the greatest value.

- PLANNING covers the activities related to identifying needs, evaluating different alternatives, selecting the right option, allocating funds, and agreeing on timing.
- ACQUISITION consists of the design and procurement of the identified physical assets or asset systems based on specific criteria and required performance standards.
- INSTALLATION AND COMMISSIONING involve the creation, building, and installation of the physical assets or asset systems and ensuring they function as per performance requirements.
- OPERATIONS cover the utilization of the physical assets to meet the strategic objectives and realize maximum value over its entire economic life.
- MAINTENANCE consists of the tactics required to ensure the physical assets perform all its functional requirements safely and at optimum costs throughout their economic lives.
- DISPOSAL involves removal of physical assets from operations and their disposal in a safe and environmentally friendly manner.

Organizational Functional Areas

Organizations are structured differently and can have many different functional areas. But the main ones dealing closely with physical assets are:

- SENIOR MANAGEMENT (SM) is responsible for defining the organization's long-term vision and mission as well as developing its strategic plan, goals, and objectives. It is also responsible for making key decisions on physical asset management initiatives and investments.
- ASSET MANAGEMENT (AM) is responsible for coordinating the activities that an organization uses to realize value from its physical assets in the delivery of its outcomes or objectives. Realization of value requires the achievement of a balance of costs, risks, and benefits over the whole lifecycle phases of the assets.
- FINANCIAL MANAGEMENT (FM)'s mandate is to ensure the necessary funding is available at the right time to secure resources at favorable prices. To make that happen, careful planning and accurate budgeting are required to avoid cash flow problems. Finance is also responsible for carrying out the economic analysis on investment proposals and looking for the best funding strategies.
- PROJECT MANAGEMENT (PM) is responsible for delivering capital projects as and when required by the organization. Its main purpose is to ensure that capital projects are managed from start to finish, and that they meet all the specified requirements and performance standards through the application of sound project management methodology.
- OPERATIONS MANAGEMENT (OM) is the core function of an organization. It is the process of transforming inputs such as raw materials, labor, machines, etc., into outputs such as goods and services as specified by customers. The essence of the operations function is to add value during the transformation process.
- MAINTENANCE MANAGEMENT (MM) is responsible for general upkeep and repair of equipment, buildings, and other physical assets. Its main purpose is to provide safe and reliable equipment and infrastructure to support operations by applying sound and proactive maintenance strategies on a regular basis.

Alignment

As previously mentioned, the organization-wide physical asset management approach is a complex system with interdependencies. To be able to navigate and manage the interdependencies, first of all alignment has to happen at all levels of the organization and across all functional areas. This is a huge and complex undertaking, but worth the effort, as it is the key to successfully operationalize an organization-wide physical asset management philosophy.

Alignment of an organization's functional areas activities with those activities pertaining to physical assets lifecycle phases is key to implement an effective organization-wide physical asset management system. It goes beyond the regular activities of managing physical assets or of implementing strategies and tactics. The concept of alignment helps transform organizations to become more effective in achieving their objectives and goals while delivering better performance in a more holistic manner. It is essential that staff in both the financial and nonfinancial functional areas work closely together with the joint objective of realizing maximum value from physical assets while delivering the agreed level of service. Alignment empowers the different functional areas of an organization to share and utilize information, to collaborate and make better decisions, to work together to meet service performance requirements, and improve achievement of the organizational objectives. Effective and efficient alignment across an organization helps break down functional silos and improve teamwork cross-functionally to deliver corporate, rather than departmental, goals and objectives. Many organizations are adopting the approach of "asset management is everyone's business" in an attempt to achieve alignment. The ISO standards on asset management also pay particular attention to alignment as a key and critical part in creating value through physical assets and encourage organizations to align their functional areas and coordinate their activities better.

Alignment is, in essence, the focal point of an organization-wide physical asset management system, but to work through it is a huge task. Identifying the functional areas and related activities for an organization is a relatively straightforward task. Identifying the activities involved in managing physical assets over their lifecycle phases is also a relatively simple task.

The challenge is in aligning, integrating, and coordinating those activities in the day-to-day management of physical assets and in the day-to-day business activities across the organization.

To tackle this challenge, we will have to break alignment into vertical and horizontal alignment and analyze the inherent difficulties.

Vertical Alignment

Vertical alignment is that clear line of sight between an organization's corporate strategic plan and the physical asset management activities to deliver value (Figure 13.1). It is the process of cascading down the organizational goals and objectives and transforming them into more detailed asset management plans and activities. It is also the process of doing bottom-up reality checks on the results delivered and gaining a better understanding of the existing challenges and constraints that need to be incorporated in decisions and in the development of strategies. Vertical top-down and bottom-up alignment typically takes place across the three different organizational levels: strategic, tactical, and operational.

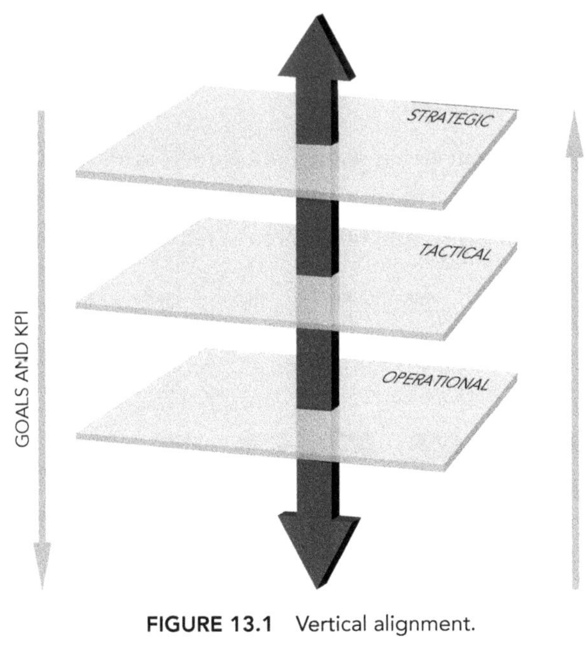

FIGURE 13.1 Vertical alignment.

Physical asset lifecycle activities occur at different points in time with various durations for different asset classes. They also take place at different levels in the organization—strategically, tactically, and operationally. It is therefore very important that there is clear and seamless connectivity between those activities across the different levels so that there is better coordination in prioritizing and working on physical asset management activities to meet the organizational strategic goals and objectives.

In the context of an organization-wide physical asset management system, this vertical alignment will be achieved through what will be called "interrelations" between the different physical asset management activities in each lifecycle phase across the three levels of an organization.

Interrelation by definition means the way in which each of two or more things is related to the other or others—in other words, how activities for physical assets are coordinated across their entire useful lives in different parts of the organization.

For example, let's say senior management decides to upgrade existing physical assets with new technologies to improve productivity. How does that interrelate with the actual implementation and commissioning of the upgrade to the actual maintenance of the new physical asset system?

Horizontal Alignment

Horizontal alignment is that strong linkage across all functional areas of an organization to achieve the organization's goals and objectives (Figure 13.2). It is the process for each functional area to refocus on its respective goals and

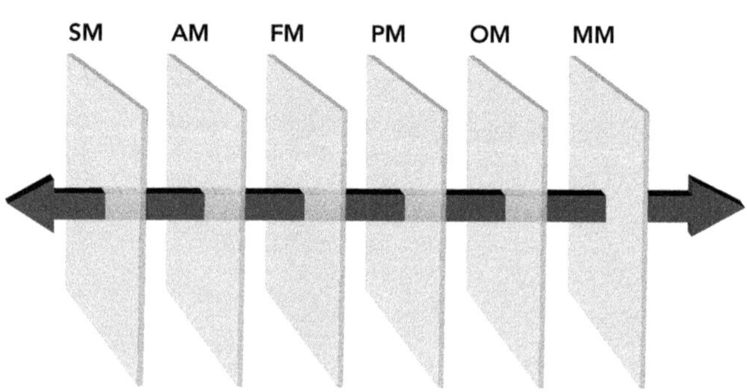

FIGURE 13.2 Horizontal alignment.

objectives to ensure these are aligned with the organizational strategic objectives. It is also the process of making sure the different functional areas work together in harmony to maximize value from physical assets.

In the context of an organization-wide physical asset management system, this horizontal alignment will be achieved through what will be called "interactions" between functional areas happening in different parts of the organization as and when required.

Interaction, by definition, means reciprocal action or influence. In other words, it means how various functional areas across the organization collaborate with each other to deliver best value from physical assets. This can mean two-way or multi-way interactions involving two or more functional areas collaborating.

Vertical and horizontal alignments, hence interrelations and interactions, are of utmost importance for an organization-wide physical asset management system.

Figure 13.3 shows a structured relationship between interrelation and interaction.

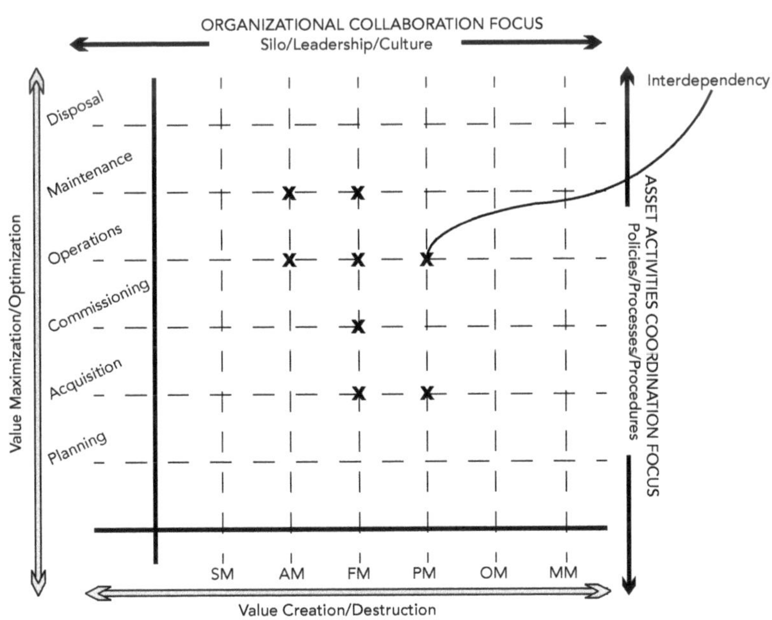

FIGURE 13.3 Interrelation and interaction in an OWPAM system.

The vertical axis illustrates the need for interrelations across all the physical asset management activities happening over the entire lifecycle phases. The focus is on policies, procedures, and processes that need to be coordinated to realize value from physical asset management activities.

The horizontal axis depicts the focus on interactions within functional areas in organizations to create (not destroy) value in a physical asset management system. The focus is on bridging the silos, instilling strong leadership to foster a culture of collaboration, and implementing the physical asset management activities.

Interrelations and interactions of physical asset management activities in different functional areas form a complex network of interdependencies. Those interdependencies take the form of different physical asset management activities in each lifecycle phase performed by certain functional areas at different levels of the organization, very often creating a chaotic situation.

In quest for vertical and horizontal alignment in an organization-wide physical asset management system, organizations have to face the challenges of navigating through the myriad of interdependencies and managing the chaos.

To start analyzing the interdependencies, it is necessary to structure the different functional areas and physical asset lifecycle phases in the form of a single matrix so as to examine the intricacies within each and every interdependency point and the effect each has on the other.

Figure 13.4 depicts an example of such a matrix, where each intersection point is an interdependency (or node) representing functional areas at specific asset lifecycle phases.

In the matrix, node A represents the interdependency of the project management functional area with the acquisition lifecycle phase. For example, the capital project team activities will have a huge influence on the design and acquisition of physical assets, which may have an impact on their performance, reliability, and lifecycle costs.

Node B in Figure 13.4 represents the interdependency of the operations functional area with the installation and commissioning lifecycle phase. An example is the involvement of operations in the installation and commissioning activities of physical assets, which may impact the level of understanding and training required to maximize performance.

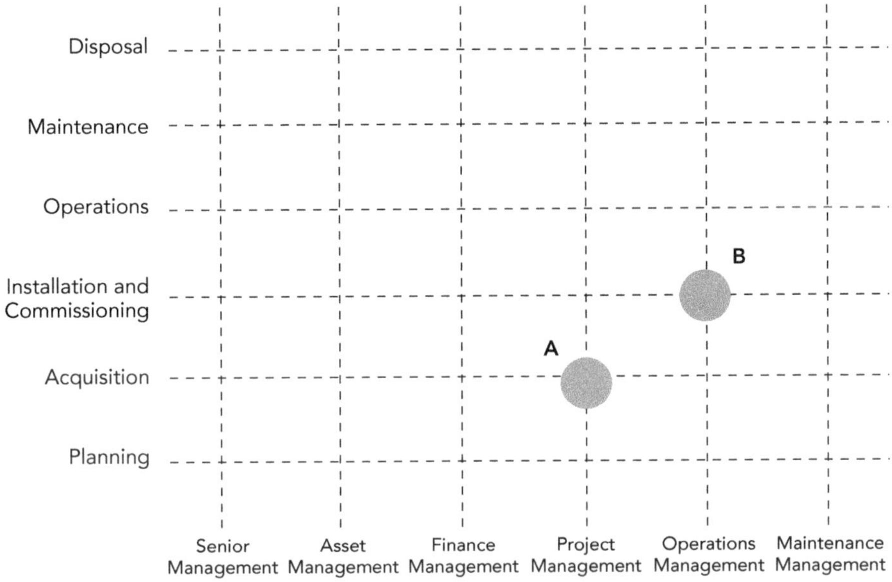

FIGURE 13.4 Physical asset management interdependency matrix.

The interdependency matrix will differ from one organization to another depending on the nature of the business, the way the organization is structured (meaning the interactions of its functional areas), and the complexity of its physical assets (meaning the interrelations of lifecycle phases).

For example, in Figure 13.5a you may have a matrix with nodes across all intersection points, meaning the existence of interdependencies between all the functional areas with all lifecycle phases.

And in Figure 13.5b you may have only a few nodes at some of the intersection points, meaning there are certain interdependencies taking place between some functional areas at specific lifecycle phases.

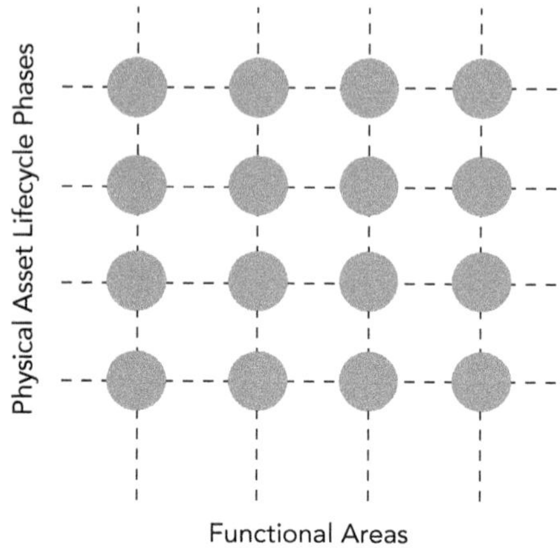

FIGURE 13.5a Matrix with nodes at all intersection points.

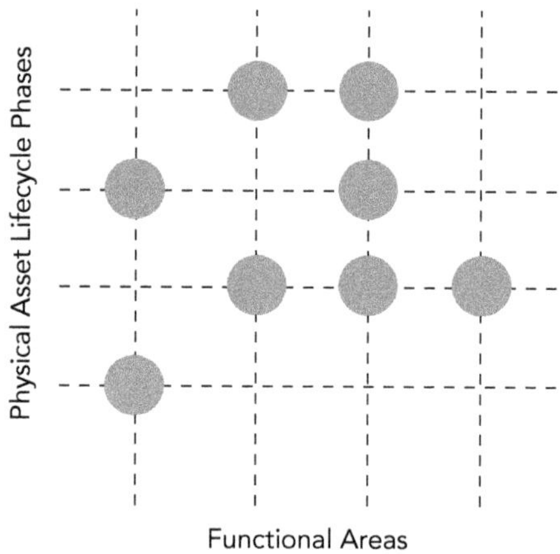

FIGURE 13.5b Matrix with nodes at some intersection points.

Now let us take a look at some of the common functional areas and asset lifecycle phases interdependencies that take place in an organization-wide physical asset management approach.

- SENIOR MANAGEMENT—PLANNING: This interdependency involves the forecasting and planning of projects around intervention on existing physical assets. It includes decisions on the prioritization of when, where, and how much to invest, thereby dictating the level of funding for operational and capital expenditures.

- SENIOR MANAGEMENT—ACQUISITION: This interdependency entails procurement policies and decisions in providing the resources and support to acquire physical assets with the right specifications, quality, price, and terms and conditions.
- SENIOR MANAGEMENT—INSTALLATION AND COMMISSIONING: This interdependency looks at the say and influence of senior management during installation and commissioning to agree on level of performance and cost at a tolerable level of risk.
- SENIOR MANAGEMENT—OPERATIONS: This interdependency involves setting goals and targets during the operations phase. This can be in the form of quantity and quality of goods and services and the strategy regarding the operation of existing physical assets.
- SENIOR MANAGEMENT—MAINTENANCE: This interdependency addresses the alignment and sponsorship of senior management to build capacity and a solid maintenance and reliability program to support operations in achieving the strategic goals.
- SENIOR MANAGEMENT—DISPOSAL: This interdependency entails a timely, consistent, and formal decision-making process from senior management based on facts to dispose of physical assets.
- ASSET MANAGEMENT—PLANNING: This entails collecting accurate data and information on asset inventory, condition, valuation, remaining life, and risks to facilitate the budget forecast and planning process.
- ASSET MANAGEMENT—ACQUISITION: This entails application of systems engineering and configuration management to ensure acquired physical assets meet required functional and performance standards.
- ASSET MANAGEMENT—INSTALLATION AND COMMISSIONING: This interdependency involves ensuring physical assets are able to meet required performance targets and business benefits with appropriate operational and maintenance plans in place.
- ASSET MANAGEMENT—OPERATIONS: In this interdependency asset management ensures physical assets are performing in compliance with set standards and requirements. Performance such as utilization rate, optimization, lifecycle costs, and efficiency is tracked, and deficiencies are recorded for future improvement.

- ASSET MANAGEMENT—MAINTENANCE: This interdependency entails implementing an up-to-date maintenance and reliability program to ensure physical assets perform as required and maintain their economic life. It includes building capacity and identifying resources and training requirements.
- ASSET MANAGEMENT—DISPOSAL: This entails an optimized and informed decision-making process to dispose of physical assets in a safe and environmentally friendly manner. It includes looking at lifecycle costs, obsolescence, failure rates, and economic factors as well as replacement options.
- FINANCIAL MANAGEMENT—PLANNING: The interdependency here is about the capital budgeting process where short-term and long-term financial plans are reviewed based on actual short-term and long-term capital investment needs forecasts.
- FINANCIAL MANAGEMENT—ACQUISITION: An example of this interdependency is the financial cash flow and disbursement needed during the acquisition process of physical assets based on specified design or performance requirements.
- FINANCIAL MANAGEMENT—INSTALLATION AND COMMISSIONING: The interdependency here is about financially supporting this physical asset lifecycle phase where the right investment is required to achieve targeted specifications and performance.
- FINANCIAL MANAGEMENT—OPERATIONS: This interdependency includes the actual accounting for new, replacement, or refurbished/betterment of physical assets in the fixed asset sub-ledger during the economic life of the asset.
- FINANCIAL MANAGEMENT—MAINTENANCE: This interdependency is about providing the funds required to maintain the physical asset in a state of good repair to meet required levels of service. It includes operational and capital expenditures required over the physical asset lifecycle.
- FINANCIAL MANAGEMENT—DISPOSAL: When physical assets reach their end of service life and need to be replaced or disposed of, finance needs to account for that in the fixed asset sub-ledger, which includes writing-off and other adjustments to the system.

- PROJECT MANAGEMENT—PLANNING: An example of this interdependency is the handing over of the capital plan to the project delivery team to execute the project as per set scope, timeline, and budget in order to achieve the required goals and benefits of the project.
- PROJECT MANAGEMENT—ACQUISITION: This interdependency consists of the project delivery team in the initiation and planning project lifecycle phase. This includes the design and procurement of assets to meet agreed-upon specifications and requirements.
- PROJECT MANAGEMENT—INSTALLATION AND COMMISSIONING: This is a very important interdependency in the physical asset lifecycle phase, as a lot of things can and usually do go wrong. It includes such things as testing, adjusting, and handing over new or refurbished assets to operations and maintenance while ensuring all specifications and performance requirements are met.
- PROJECT MANAGEMENT—OPERATIONS: This interdependency talks about the involvement of the project team in the operations process throughout the entire lifecycle of the project. It entails understanding and including operational performance requirements as part of the project benefits.
- PROJECT MANAGEMENT—MAINTENANCE: This interdependency deals with the inclusion of physical asset maintenance and reliability requirements throughout the project lifecycle phases. It entails involving maintenance stakeholders especially at the design, installation, and commissioning stages.
- PROJECT MANAGEMENT—DISPOSAL: This interdependency is about the requirements and procedures to follow during the project delivery process when it's time to dispose of a physical asset and its components.
- OPERATIONS MANAGEMENT—PLANNING: This interdependency includes needs identification, operational planning, and capital planning to meet operational needs (e.g., new products and services will require different or higher levels of services which will trigger renewal/upgrade/replacement of assets).
- OPERATIONS MANAGEMENT—ACQUISITION: This interdependency is where operations gets involved with procurement, and engi-

neering and external suppliers to define and agree on asset specifications, configuration, and performance standards.

- OPERATIONS MANAGEMENT—INSTALLATION AND COMMISSIONING: Operations and the installation and commissioning asset lifecycle phase is very important to ensure assets are installed/constructed as per required specifications and configuration and adjusted to meet the agreed-upon operational performance requirements.
- OPERATIONS MANAGEMENT—OPERATIONS: This interdependency includes the development and implementation of operating procedures, operational plans, and controls to maximize performance and realize the value from the assets.
- OPERATIONS MANAGEMENT—MAINTENANCE: This interdependency is the longest, and is very crucial for operations to define, track, and control asset performance standards to ensure maintenance activities are able to maintain the desired level of performance.
- OPERATIONS MANAGEMENT—DISPOSAL: This interdependency is about operations identifying any asset that is surplus, is obsolete, or no longer meets operational requirements and ensuring action is taken to dispose of.
- MAINTENANCE MANAGEMENT—PLANNING: This interdependency is the involvement of maintenance in the operational and capital planning process to identify any minor and major asset repair, renewal, refurbishment, or replacement required as well as forecast estimates and timelines for execution.
- MAINTENANCE MANAGEMENT—ACQUISITION: This interdependency includes the involvement of the maintenance function in the design and procurement phase of assets to ensure they are as per defined standards and technical specifications.
- MAINTENANCE MANAGEMENT—INSTALLATION AND COMMISSIONING: This interdependency, one of the most overlooked in many organizations, is where the maintenance function gets involved and fully participates in the installation and commissioning phase of assets in order to better understand their operational and maintenance issues and get trained before the assets are handed over to maintenance.

- MAINTENANCE MANAGEMENT—OPERATIONS: This interdependency includes the maintenance function fully understanding the operational requirements and constraints and the asset performance standards needed by operations to deliver products and services.
- MAINTENANCE MANAGEMENT—MAINTENANCE: This interdependency is about the development and implementation of maintenance tactics to meet regulatory requirements and operational performance standards while realizing maximum value from the assets.
- MAINTENANCE MANAGEMENT—DISPOSAL: This interdependency includes the maintenance function ownership, identification of redundant assets, and the decision-making rationale to dispose of the assets in a safe and environmentally responsible manner.

By now we know that there is a complex network of interdependencies present across organizations to deliver value from physical assets. We have to also understand that the interdependencies have varying degrees of importance in different organizations depending on the operating context.

When analyzing the interdependencies between functional areas and asset lifecycle phases in an organization-wide system using the physical asset management interdependency matrix, there are three very important factors to consider:

1. The nodes may be connected to each other vertically to achieve that vertical alignment. These are the interrelation linkages between the different physical asset management activities happening at different lifecycle phases across the different levels of the organization. To achieve the organizational strategic goals and objectives and to realize maximum value from physical assets, managing those linkages seamlessly and in an interrelated manner over the entire lifecycle phases is very important.

   In Figure 13.6, we can see that there are strong linkages between the acquisition phase and the installation and commissioning and maintenance phases within the project management function. For example, the design and acquisition of physical assets by project management can largely impact the way they are going to be installed and commissioned as well as the way they are going to be maintained to achieve best performance. On the other hand, the linkage with the planning lifecycle phase can be weak.

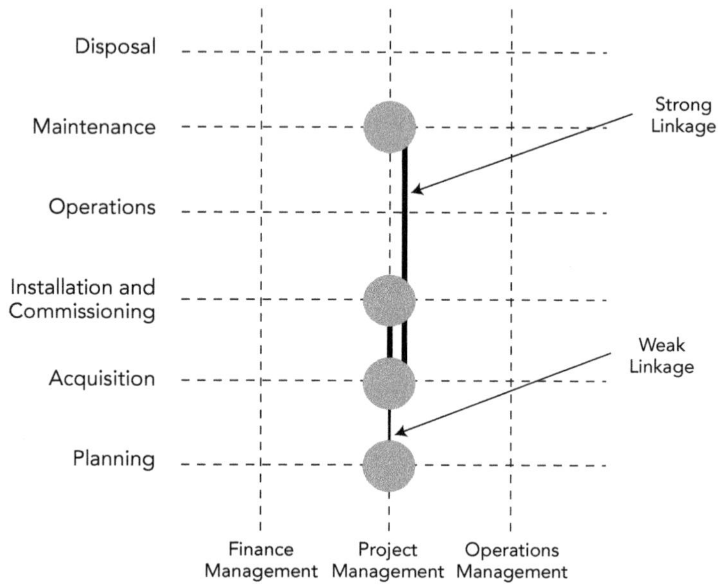

FIGURE 13.6 Interrelation nodes linkages.

Depending on the nature of the organization, its complexity, and how asset-intensive it is, the linkages can vary from weak to strong from one lifecycle phase to another within each functional area.

2. The nodes may also be connected to each other horizontally to achieve the horizontal alignment. These are the interaction linkages between the different functional areas' physical asset management activities across the organization to achieve the organizational strategic goals and objectives. Collaboration and close interactions between functional areas are very important to realize maximum value from physical assets.

   In Figure 13.7 we can see that there are strong linkages of the operations function at the installation and commissioning phase with the maintenance and project management functions. For example, operations can have a major say on the physical assets' operating and performance standards at the Installation and commissioning phase that can largely be influenced by how project management carries out the installation and how Maintenance understands the maintenance requirements to meet those standards. On the other hand, the linkage of operations management with finance can be weak at the installation and commissioning phase.

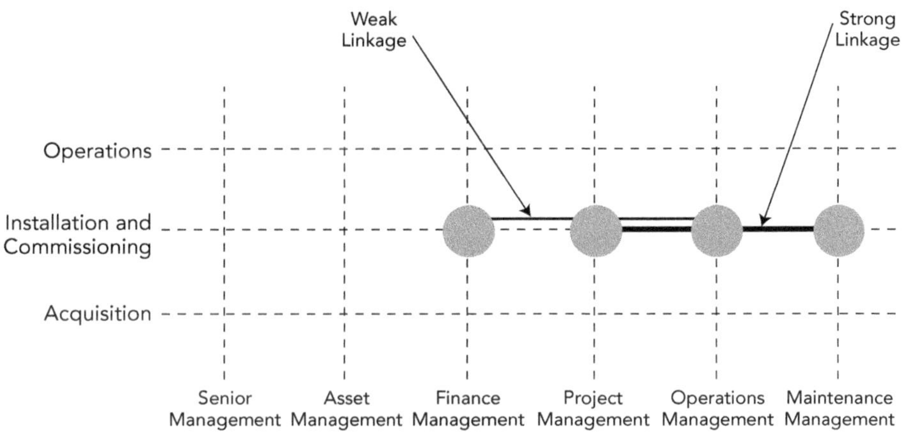

FIGURE 13.7 Interaction nodes linkages.

Depending on the culture and structure of the organization and the business complexity, the linkages can vary from weak to strong from one functional area to another within each lifecycle phase.

3. The third and final factor that is important when analyzing interdependencies in a physical asset management interdependency matrix is the size of the node. The size of the node determines the degree of interdependency taking place between the functional area and corresponding physical asset lifecycle phase. This can be a representation of the quantity and quality of physical asset management activities taking place or necessary to take place within that node.

To illustrate this concept, we will use four sizes of nodes as shown in Figure 13.8. These are examples to demonstrate the different degrees of interdependency that could exist within a physical asset management interdependency matrix.

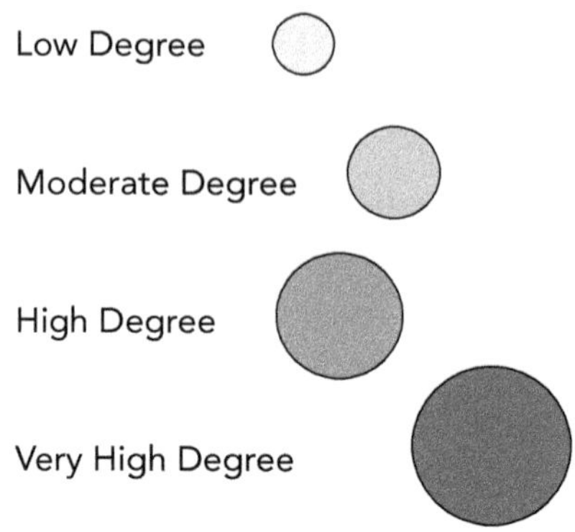

FIGURE 13.8 Degrees of interdependency.

This time we will take the example of the finance management functional area and the maintenance lifecycle phase node to illustrate the degree of interdependency scenario.

Figure 13.9 shows that there is a moderate degree of interdependency existing between the finance management functional area and the maintenance lifecycle phase. This node can represent simple activities in finance where a simple asset sub-ledger is kept and basic maintenance costs are collected.

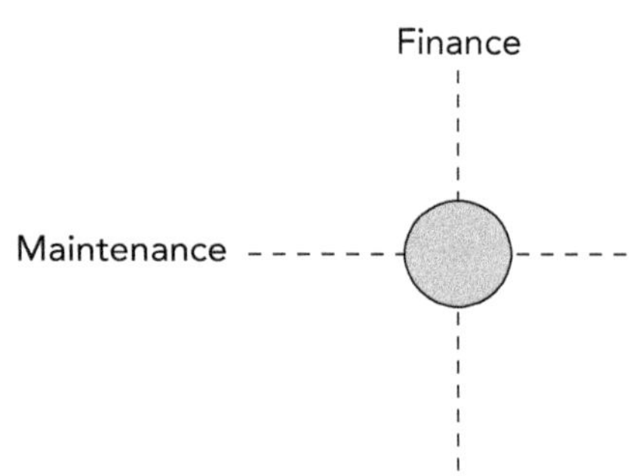

FIGURE 13.9 Example of a moderate degree of interdependency.

Figure 13.10, on the other hand, shows a very high degree of interdependency between the finance management functional area and the maintenance lifecycle phase. This can be in the form of significant high-quality interactions and activities between finance and maintenance such as activity-based costing, compiling, tracking, and analyzing physical assets' lifecycle costs; or it might be maintaining and updating the asset sub-ledger to reflect asset modifications and changes.

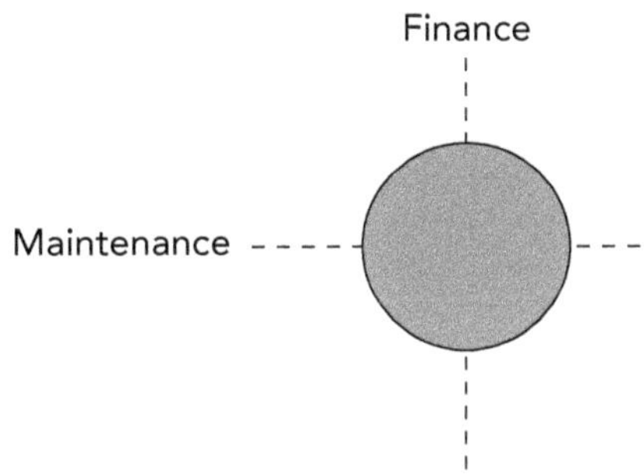

FIGURE 13.10 Example of a very high degree of interdependency.

It must be made clear that not all organizations need to have such a high degree of interdependency between finance and maintenance as shown in Figure 13.10. Some organizations may be just fine with a moderate degree of interdependency as long as they remain in compliance and all their requirements are met.

So how would organizations know what degree of interdependency they would need not only for that node but for all the other nodes in the physical asset management interdependency matrix?

To answer this question let us look at what organizations should do about the complex network of interdependencies by introducing the physical asset management bubble chart.

Physical Asset Management Bubble Chart

Earlier we introduced the physical asset management interdependency matrix, where the network of interdependencies between functional areas and physical asset lifecycle phases for an organization-wide system is plotted. When each node is evaluated to determine the degree of interdependency that exists, the matrix will produce a bubble chart.

Figure 13.11 shows an example of a completed physical asset management bubble chart.

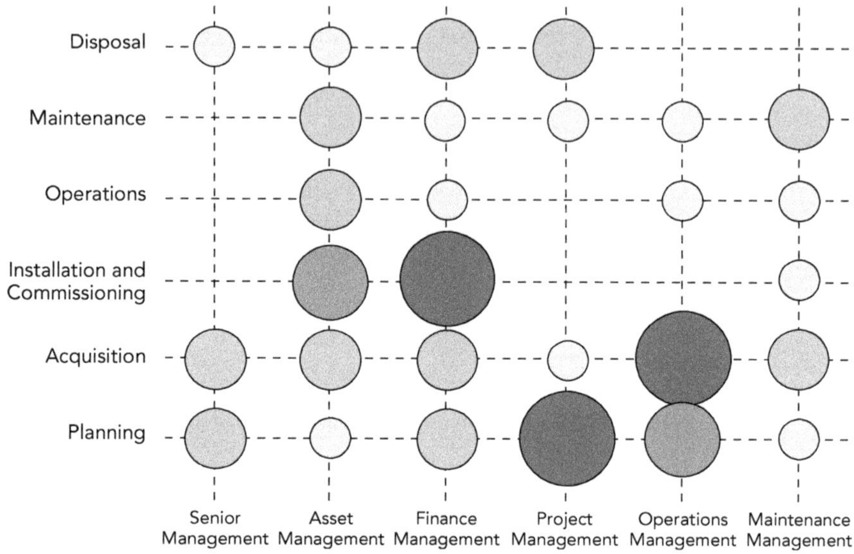

FIGURE 13.11 Physical asset management bubble chart.

The distribution and size of the bubbles give an obvious indication of the areas in the organization-wide physical asset management system where interdependencies are of greater or lesser significance. For example, in Figure 13.11, looking vertically up there are many interdependencies within the asset management and finance management functions across all phases of the physical asset lifecycle. This means that the activities are closely interrelated and there is good flow of information required from one phase to the other in those functional areas.

On the other hand, looking horizontally across there are many interdependencies within the planning and acquisition lifecycle phases across all functional areas of the organization. This means that close collaboration and interaction among all the different functional areas in the planning and acquisition phases are of great significance.

At this point it is worth mentioning that these are only estimates meant to provide a relative mapping of the degree of interdependencies existing within an organization's physical asset management system for each functional area with respect to every physical asset lifecycle phase.

It goes without saying that organizations will eventually have different physical asset management bubble charts with different areas of concentration and degrees of interdependencies, depending on how asset-intensive they are and the complexity of their physical assets. For example, one organization may have a higher degree of interdependency in the project management and installation and commissioning node than another organization due to the nature of the organization's business and operating context. An organization can draw its physical asset management bubble chart to understand its prevailing *as-is* situation. The chart will shed some light on the current status of its organization-wide physical asset management system and can be used to identify:

- The existence and concentration of interdependencies
- The current degree of interdependencies across all nodes
- The level of interaction prevailing across functional areas
- The level of interrelation throughout all lifecycle phases

From there the organization can analyze any gaps and/or overlaps that may exist to get to a desired *to-be* situation.

- GAPS where there is no bubble present while there should have been, or where the bubble is small in the *as-is* chart while it should have been bigger, meaning potential sign of gaps
- OVERLAPS where many bubbles are present and overlapping in a lifecycle phase, or where the degree of interdependency is too high, indicating potential duplication of, or too much focus on, certain activities

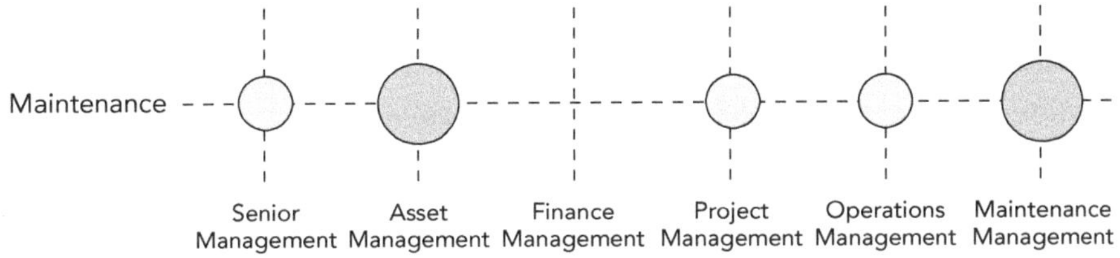

FIGURE 13.12 Example of potential gaps.

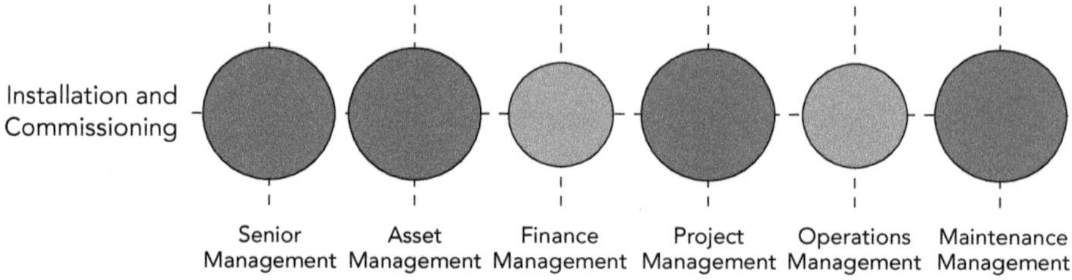

FIGURE 13.13 Example of potential overlaps/duplications.

When organizations know their *as-is* situation, they can decide and draw their *to-be* status as illustrated in Figures 13.14a and 13.14b.

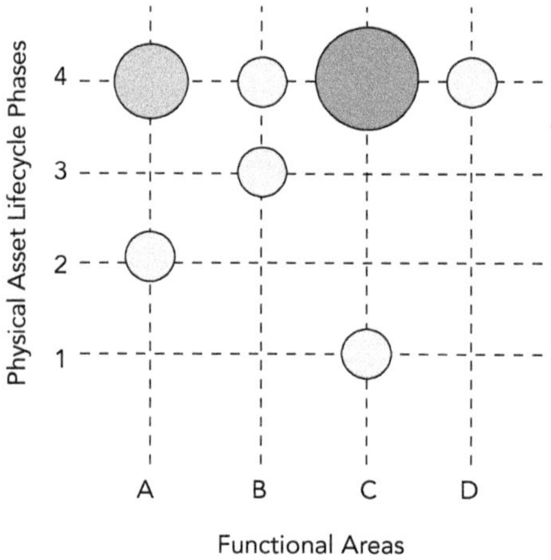

FIGURE 13.14a As-is situation.

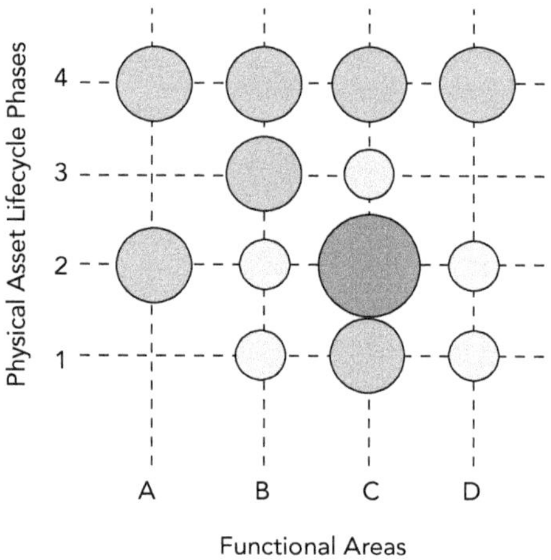

FIGURE 13.14b To-be situation.

Once the to-be status is approved, organizations can start developing roadmaps and plans to act on the identified gaps and overlaps, to move from the as-is to the to-be status.

Before going any further, there is a twist to the complexity of the organization-wide physical asset management system matrix that adds another dimension to the complex network of interdependencies.

Physical Asset Management System of Interdependencies

So far, we have seen that we can build an intricate matrix/bubble chart to show the network of interdependencies and the various degrees of interdependencies existing in different functional areas of an organization over the entire lifecycle of physical assets. In an organization-wide physical asset management system, this helps greatly in the process of achieving horizontal alignment and better collaboration to maximize value from physical assets.

The physical asset management bubble chart is, however, only a two-dimensional representation of the network of interdependencies and only partly illustrates the full spectrum of interrelations and interactions taking place in organizations. To get a better idea of an organization-wide physical asset management system and the full extent of its network of interdependencies, we need to look at an additional dimension.

If you look closely in organizations, you'll see that physical asset management activities (by functional areas and within different lifecycle phases) take place at different levels of the hierarchy as well (i.e., at the strategic, tactical, and operational levels). As explained earlier, this set of interrelations and interactions is very important to achieve vertical alignment.

When we combine horizontal alignment with vertical alignment, another equally important dimension of intricate interdependencies between activities is uncovered.

From Figure 13.15 we can see that interdependencies at the operational level can influence and translate into interdependency at the tactical level, which in turn becomes part of interdependency at the strategic level—also impacted by other interdependencies as well. This is a bottom-up scenario. Now we can have a similar scenario if we take a top-down perspective where interdependencies proliferate from the strategic to the operational level of the organization. Either way it creates a complex three-dimensional network of interdependencies highly influenced by the quality of interactions between functional areas, by the interrelations through lifecycle phases, and by alignment across different levels of the organization.

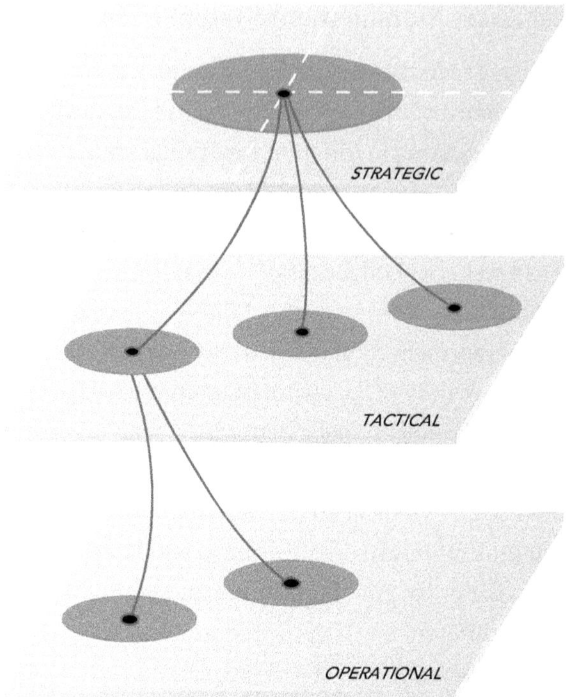

FIGURE 13.15 Another dimension of interdependency.

- An example is illustrated in Figure 13.16. Consider the node of maintenance management functional area and operations lifecycle phase in an organization at the strategic level. Maintenance is unable to meet the operational objectives and market demands, hence impacting the market share of the organization. At the tactical level this can be a combination of three nodes consisting of poor project management in delivering the project benefits, substandard operating procedures, and inadequate maintenance tactics in place. The node project management and operations lifecycle phase at the tactical level may have interdependencies at the operational level from a combination of lack of understanding of operational requirements during project design and poor proactive maintenance completion rate. This is illustrated in Figure 13.16.

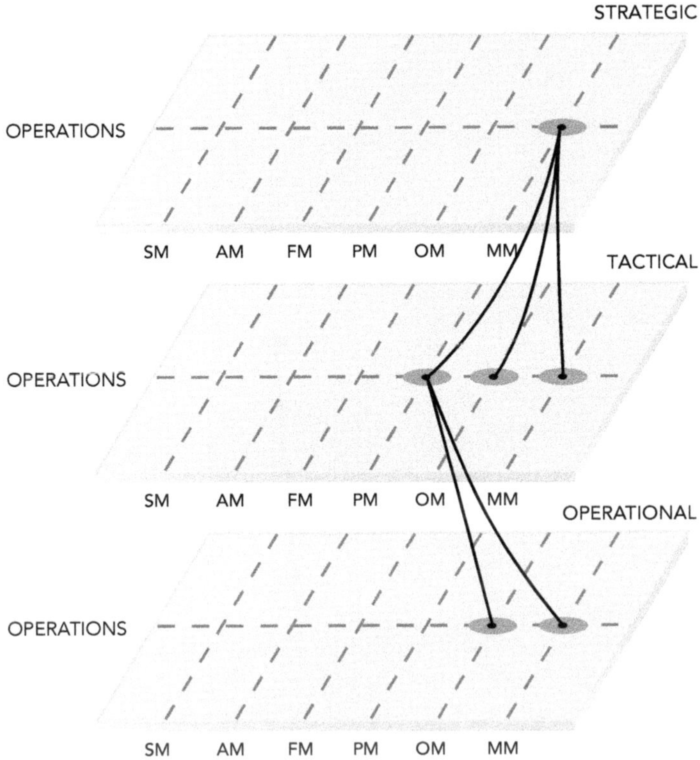

FIGURE 13.16 Example of three-dimensional interdependencies.

Now interdependencies can also be diagonal (i.e., they don't need to add and roll up to a higher level of interdependency within the same lifecycle phase or functional area).

For example, the node project management and operations lifecycle phase at the tactical level for poor project management in delivering the project benefits may have interdependencies in the installation and commissioning lifecycle phase as well as at the operational level (Figure 13.17).

This explains the theory behind the complex network of interdependencies existing in an organization-wide physical asset management system. Technically it does not need to be drawn, but you need to be aware of and acknowledge the existence of such a network of interdependencies. Only then will you be able to navigate through and tackle them.

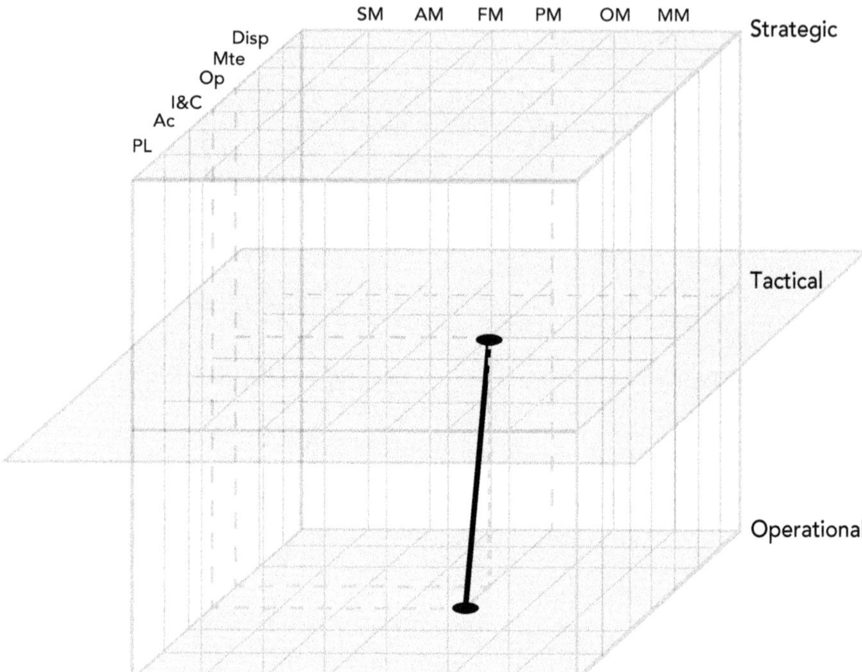

FIGURE 13.17 Example of diagonal interdependency.

Figure 13.18 summarizes the three-dimensional representation of the physical asset management interdependency chart. It represents the complete network of interdependencies and encapsulates:

- All functional areas
- All physical asset lifecycle phases
- All nodes at different degrees of interdependencies
- All interdependencies at all levels

The concept is similar to the physical asset management bubble chart, except that now the interdependencies are analyzed in a three-dimensional matrix. Nodes of different degrees of interdependency and varying linkages are suspended and connected to each other at different levels. Any deficiency at a node is transmitted through the linkages to other nodes in the system. Depending on the size of the node (degree of interdependency), the effect on

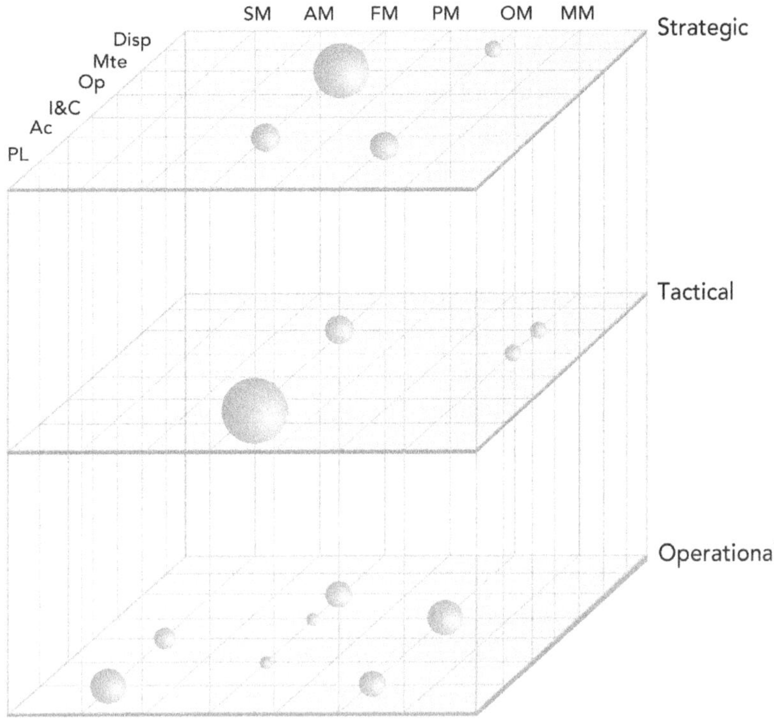

FIGURE 13.18 Three-dimensional physical asset management bubble chart.

the other nodes can be significant. Similarly, depending on the strength of the linkages, one node can be affected more than another in the entire system.

This network of interdependencies is reflected in an organization-wide physical asset management system where a deficiency in one process/activity at any level in the organization can have a corresponding impact elsewhere. The magnitude and seriousness of the impact will largely depend on the degrees of interdependency and strength of linkages with other processes/activities in other lifecycle phases and functional areas.

Every organization will have a different three-dimensional physical asset management bubble chart to represent its physical asset management system's network of interdependencies. The complexity of the network will depend on the nature of the business and complexity of its physical assets. Some organizations may be more impacted at one level than another, while other organizations may find that the focus on certain functional areas may vary. For example, there can be significant differences between the network of interdependencies of a food manufacturing facility which is very asset-intensive and a university facility complex with larger less-intensive assets. There may be notable focus on the operations and maintenance management activities in one case and more project management activities in the other. There may be more processes involved in the operations lifecycle phase in one situation and more planning and acquisition processes in another.

Even though organizations will apply the same organization-wide philosophy to build their physical asset management system, their network of interdependency matrices will be completely different from each other. It is extremely rare to find two similar network interdependency matrices, because the variables defining and determining each matrix are many and complex.

The organization-wide physical asset management system of interdependencies is a useful tool to provide the organization with a structured way of understanding the complexity of its processes and activities and the impact they may have on each other at different levels in the organization.

Darren stops and exhales. He looks at his watch and notices that it is almost two o'clock. He thinks to himself that Debbie will be back from shopping at any moment. He better take a break and start making some tea before she comes. Deep down there is something still worrying him. He is thinking about how to take the concept of the organization-wide physical asset management system of interdependencies to the next level and apply it in a real organization as a proof of concept.

CHAPTER 14

OWPAM SYSTEM OF INTERDEPENDENCIES

It has been quite a busy few weeks for Darren as he prepares for the start of the new academic year. Summer has gone by quickly, and Darren has taken full advantage of the gorgeous weather to train outside for his upcoming race. Darren's summer also consisted of a few local family excursions. For the next few weeks his main focus is to maintain a good balance between training and preparing for the start of the new school year. Darren understands the importance of meticulously training more lightly his training regime to get to race day in the best form possible, both physically and mentally.

After reviewing his email, Darren checks his electronic calendar and, as usual, starts updating his training schedule. Suddenly, an incoming email pops up on his screen. It's from Jerry.

Jerry has been quiet lately; the last time they met was at the seminar. Darren has not heard from him at all since then, which is probably a good sign. He knows Jerry will always reach out to him when he needs help or would like to discuss something.

He opens the email wondering what it is going to be about.

Hello Sir,

I hope this email finds you well. Here at Mouroum Inc. things are well, and we are making good progress on the asset management front. Yesterday I had a meeting with our Senior Leadership Team to provide an update on the asset management project. They were satisfied with our progress, but they were worried about the next steps. They said that, from their experience, they have seen many AM programs fail or stall after a few years, and that organizations fail to operationalize their AM plans to produce results.

They asked me what to do, as they want to do it right the first time rather than embarking on such a journey without a clear understanding of what this will lead them to.

I remember what was covered in your seminar, and I mentioned to them about the OWPAM approach. Mr. Vuhdam Jeragvika was impressed by the approach, but he wants a greater understanding of the matter.

He asked me to contact you to see if you will be willing to come by and explain the OWPAM approach to his team. He is willing to pay you for your time. He thinks this will be of great benefit to his team and to Mouroum Inc.

Please let me know what you think and what your availability is in the next coming weeks.

At first Darren is not compelled by the offer because of the three-week notice. The proposed window of time is before his race, and he wants to remain focused on his race with no distractions. Though after a few minutes of reflection, he changes his mind. For some time now he has actually been on the lookout for such an opportunity to take the OWPAM concept a step further and apply it as a proof of concept.

"This seems to be the perfect opportunity, and who best can we have to operationalize it other than Jerry!" Darren says to himself.

With a large grin on his face, he starts typing his response to Jerry.

On a lovely Thursday morning, Darren parks his SUV in the Mouroum Inc. parking lot in a "Reserved for Visitors" spot. He peeks at his watch and is glad he made it on time after going through the painful morning traffic. He still has five minutes, and he hastily heads toward the main entrance. The head office is a majestic building and really represents the prestige and pride of the company. Darren announces himself at reception and waits in the lounge while looking at the trophies and plaques that Mouroum Inc. has won over the years.

"Good morning, sir!" Jerry suddenly says as he runs down the stairs and walks toward him.

"Good morning," Darren responds as he shakes Jerry's hand vigorously. "Good to see you, Jerry, after such a long time. How has it been?"

"Pretty good," replies Jerry with a big smile. "I am very delighted that you have accepted our invitation to come by to talk to our senior executives. They are expecting you and are keen to hear from you."

The executive boardroom is on the top floor of the building as you would expect.

In the room, three ladies and three gentlemen are already there, some on their phones and others on their laptops.

Jerry quickly introduces Darren to each one of them.

First he meets the COO Robert Veristrik, who is Jerry's boss, a tall and chubby guy in his late fifties with a short haircut and stern face. He greets Darren in a deep voice and with a powerful handshake.

Next is Mary Maswiv, the CFO, who is of medium height and with gray shoulder-length hair. She is elegantly dressed in a maroon suit and greets Darren with a big smile. Shika Keyan, a tall and thin lady who must be in her forties, is the CHRO. She has a very pleasant look with curly brown hair and red lipstick and is dressed in a black dress.

"Nice to meet you, Mr. Ordlan!" she says with a large smile while extending her hand for a handshake.

Pravin Dimitrin is the CIO of Mouroum Inc. He is relatively short with a medium build and wearing a dark gray blazer over a white shirt. He switches his focus from his laptop, removes his big square glasses, and looks up to greet Darren in a gentle voice.

The CMO, Vinn Matis, is completely different. He is a very outgoing tall gentleman with an athletic build, most probably in his mid-forties. He greets Darren with a large smile and a firm handshake.

"Welcome, Darren. Great to have you here," Vinn says.

And finally Darren meets with Nastia Niaran, executive assistant to the CEO. She is a young lady in her early thirties, neatly dressed and well organized, at the corner of the room.

The CEO, Vuhdam Jeragvika, is not here yet, but his voice could be heard in the corridor.

Darren sets up his laptop and grabs a glass of water. He then takes his seat next to Jerry while they all wait for the Chief Executive.

"Good morning, everyone," says Vuhdam as he enters the room a tad out of breath.

"I guess you all have introduced yourselves to Darren," he continues as he walks to greet Darren with a big smile and powerful handshake.

"It's great to have you here, Darren," he says.

Vuhdam then takes his usual seat at the head of the table and stares at Nastia, who gives him a nod.

Vuhdam Jeragvika is a medium-built gentleman who must be in his early fifties, but looks fit. He is elegantly dressed in a navy blue suit and has neatly combed salt and pepper hair. He is the typical leader who has a charismatic presence in the boardroom. You can feel the respect and anticipation for the guy as soon as he enters the room.

"OK. Let's get going then," he says in a decisive tone as he clears his throat.

"First of all, on behalf of the whole Mouroum Inc. team, welcome, Mr. Darren Ordlan. We are happy to have you here to enlighten us on the path to operationalizing physical asset management. As Jerry may have mentioned to you, we have a total of 8.5 billion dollars' worth of physical assets in our six plants. We want to make sure we have a good system in place to wisely manage our physical assets in a consistent manner as well as making the best decisions when required."

"Jerry has done a fantastic job so far, but I have the impression that we have stalled. From what I am hearing, there is some kind of chaos and confusion."

He stops, takes a deep breath, and continues, "The feedback I'm getting is that there is a feeling of frustration and fear of the unknown. Is there anything we should be doing differently?"

Darren hesitates for a few seconds as lots of things race through his mind. He is thinking of where to start, how much detail they want to hear, and, most importantly, how much time they have. It is no secret that, at this level, you have to cut to the chase to convey the key messages without losing the audience.

"You are not doing anything wrong," Darren starts. "I would even say that you are doing well compared with many other organizations for two reasons. Firstly, for what you have accomplished so far in terms of the policies and strategies developed; and second, because physical asset management is being discussed at this level of the organization. Seeing that you are all dedicating your precious time to listen to me to understand how to best approach physical asset management demonstrates in itself great leadership and commitment."

Darren pulls up his first slide on the screen (Figure 14.1) and continues: "And leadership is one of the fundamentals of asset management. Physical asset management is based on a set of fundamentals, which provide the critical focus required to develop and implement an effective organization-wide physical asset management system."

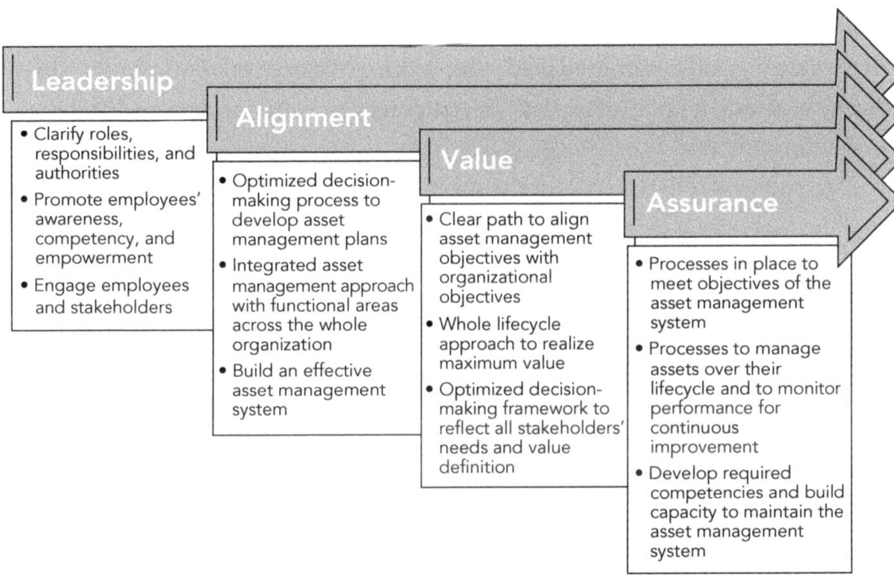

FIGURE 14.1 Fundamentals of asset management.

Leadership

Darren explains, "Implementing physical asset management is not only about implementing the strategies, policies, and plans; it's much more than that. It is a transformation of your organization; it is about moving it from the chaos that you rightly mentioned into clarity. And this will take time, lots of time."

"What I have noticed over the years is that organizations tend to start at the wrong spot and try to reach the finish line quickly. But that usually does not last. At some point they will stall; they will get stuck with frustration."

"Why is that so?" Vuhdam asks.

"Well, simply because we have not taught people what to do nor taken the time to explain why," Darren replies. "As I mentioned, it's great to see leadership commitment at this level. The next step is to ensure all staff, at all levels of the organization, have the same commitment by making them aware of asset management and its benefits. This is the first rudimentary thing to do, and most organizations fail at it because they assume that everyone knows."

"Physical asset management is an organization-wide system with a complex network of interdependencies that impacts the whole organization. A key factor is to have a strong and committed leadership team to manage this complexity. This is done by breaking it up into its parts, or structures, to better understand and manage the components."

Darren pauses, looks around the room, and then smiles. "Unfortunately," he says, "the facts show that this level of understanding is not always present at the senior management level in most organizations."

Jerry is the first to feel a little uneasy with what Darren just said. Everybody else in the room looks a little skeptical and even agitated about the statement, but nobody says anything except for Vuhdam.

"Interesting point, Darren," he says. "I would like to hear more about that and what you actually mean."

"From my experience, I believe that physical assets are as important as the human, financial, or data assets used to produce goods and deliver services. They have to be planned and managed with a strategic approach similar to what you use for all other assets. What we find is that physical asset management is not adequately represented, discussed, and understood at the senior management level."

"I was explaining this to Jerry the other day regarding a study that was conducted on a hundred major companies globally. Based on the numbers in their financial statements, they manage on average close to 18 billion dollars of fixed assets, and they don't have someone at the executive level to 'own and be responsible for' these fixed assets similar to those responsible for other types of assets in the organization."

"But we do have our COO, Robert Veristrik, to bring up the operations and physical assets perspectives in our meetings," replies Vuhdam.

"Good point," replies Darren. "I would say 'yes' and 'no.' There can be some conflict of interest here, as the COO may be torn between meeting operational targets and making best optimal decisions for physical assets. There may also be conflicts with short-term gains versus long-term benefits."

Robert Veristrik, seated facing Darren, seems a little irritated with Darren's comments. A bit heatedly, he asks, "What do you mean then, that we should have a Chief Physical Asset Officer or something similar?"

"Not really," Darren responds. "That's not what I am trying to say. Yes you can have such a position, if that is what the organizational structure requires, but you can also do it differently."

"How?" Mary Maswiv, the CFO, asks in a calm voice.

"Well," Darren replies, "leaders do not need a title to function. If you provide an individual the proper support and empowerment, you can achieve the same goals. What is important is that, at the senior management level and anywhere else in the organization, there is someone who can advocate for the benefits of physical assets and sponsor the management philosophy."

"Physical asset management should have the same visibility and importance as all other assets at the senior management strategic level, not only at the tactical and operational levels. A clear line of sight must exist throughout all levels in the organization to evaluate the impact of physical asset-related decisions. Senior management must understand and commit to the short-term and long-term implications of good physical asset management practices on the overall organizational strategic objectives."

"Does that mean that Jerry's position should be at the corporate level?" Vuhdam asks, with his hands behind his head.

"It could be a good approach," Darren says as he glances at Jerry, who has been quiet all this time.

"Can you tell us about the leadership qualities or aspects that we should have in place for this particular situation?" Shika Keyan, the CHRO, asks.

"Great question!" Darren exclaims with a smile. "One thing we have to remember is that implementing an organization-wide physical asset management system is not a mere task of implementing an information management system, for example. It is much, much more than that."

He pauses and takes a sip of water from his glass. Then continuing on, he says: "It is a huge transformational undertaking that will require focusing on the interactions and dynamics, or patterns, to understand the network of interdependencies involved in a complex system. What this transformation requires is a true revolution in our way of thinking and behaving. We cannot get away with simply changing our outward behavior or our corporate image. We are talking about a change that deeply affects the way each one of us sees the world, our view of reality, our way of thinking."

"It looks like we are talking about a strong change agent," Vuhdam quietly interjects.

"Rightly so," Darren replies. "Leadership must be in place with the main task of creating awareness and acknowledgment. This is key to building the culture and accomplishing the transformation process within the organization in order to move from a silo-based management style into one of system optimization. Leaders across all levels of the organization must focus on changing their management approach to become a more interconnected system, where cooperation is more important than competition, where everyone strives for win-win situations, and where the organization can achieve its maximum potential."

Darren looks around the room for any reaction. Everyone seems to be listening, so he continues: "It means shifting from the old-fashioned command and control style of leadership to a new leadership style where competition is replaced by collaboration, where performance is managed systematically, where teamwork is fostered rather than the ranking of individuals."

Darren quickly searches for another slide on his laptop and brings up the definition of asset management. "The essence of physical asset management is coordinating all the activities related to physical assets!" he adds. "It's about building the interrelations and interactions across the whole organization to create an interconnected organization and provide everyone with a clear vision of a better system."

"OK," says Vuhdam. "Let's assume we have the leadership in place, which I believe we are on the right track with by having Jerry on board. He has done a great job in developing the asset management policy and the strategic asset management plan that he presented to us. How do we move on from there?"

"The next obvious thing is for everyone in the organization to face reality!" Darren responds.

There is a look of surprise on everyone's face in the boardroom.

Value

"Allow me to explain," Darren quickly adds after realizing the situation. "I will keep reminding you that physical asset management is an organization-wide system with a complex network of interdependencies. Managing those interdependencies and the physical asset management system is what will help achieve best value from physical assets. Physical assets exist to provide value to the organization and its stakeholders. Value can be tangible or intangible, financial or nonfinancial, and in the form of goods or services. Furthermore, physical asset management does not focus on the physical asset itself but rather on the value provided to the organization."

After a short pause to clear his throat, Darren points at the huge company logo on the wall and notes: "Mouroum Inc. invests money to buy physical assets, invests money to hire people to design, build, operate, and maintain those physical assets, and invests money to implement software systems to collect data on the same physical assets to make the best decisions at the right time. So there must be no doubt about the fact that physical assets are an integral part of the whole process of creating and preserving value."

Switching to the next slide (Figure 14.2), he poses some questions to executives. "The reality I want you to face is: Do you know what realizing value from physical assets means for Mouroum Inc.? Has anyone ever sat down and documented what value means in your organization, in each department, and for your stakeholders? Does everybody understand how value is created and how each and every one contributes to it?"

Darren stares at everyone in the room, and as expected, Vuhdam replies: "The way I see it," he says, "is that value at Mouroum Inc. is represented by our set of corporate objectives that are translated to sets of goals and objectives for each department. We track those objectives through different performance measures."

"I agree. This is what most organizations do," Darren says. He leans on the table and adds, "Let's think about this for a moment. Are we sure we are meeting

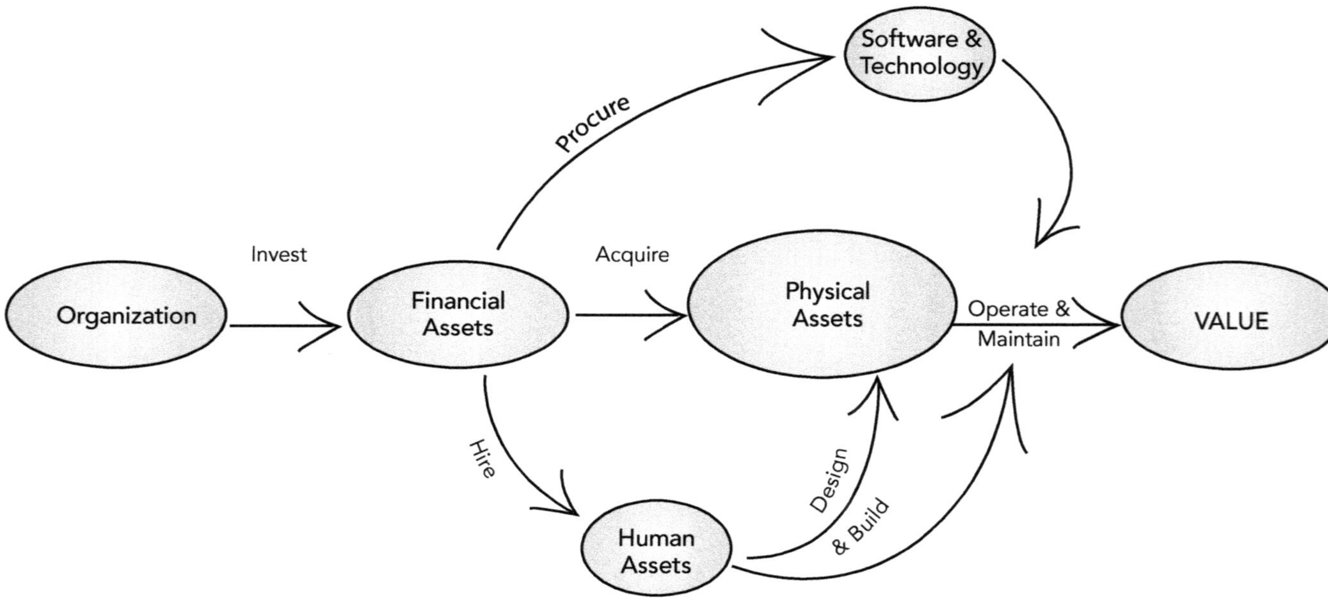

FIGURE 14.2 Value creation process.

the objectives of our customers and stakeholders? Are we sure we are creating or adding value rather than destroying value in the process of achieving these objectives? Are we sure we are realizing maximum value from the physical assets?"

"In many organizations, there are conflicting priorities and numerous constraints where decisions are made in silos, resulting in missed opportunities. In a physical asset management system with a complex network of interdependencies, several stakeholders are involved over the whole lifecycle phases of physical assets. Hence, silo decision making and missed opportunities are bound to happen. That is the reason why better-coordinated activities are required to realize maximum value from physical assets. So how do we get this complex system to perform for Mouroum Inc., to realize maximum value from your investments and, in this case, your physical assets?"

Darren pauses to make sure everybody is still with him, and then he continues: I propose to take a few distinctive approaches over and above developing and implementing the asset management strategies, policies, and plans. This will assist in realizing maximum value from physical assets in an organization-wide physical asset management complex system."

Darren flips to another slide on the screen (Figure 14.3).

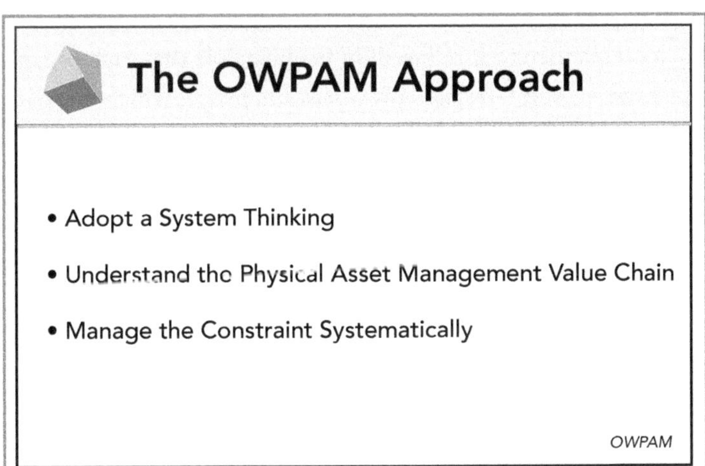

FIGURE 14.3 The OWPAM approach.

Commenting on the slide, he says, "This may look like a complex undertaking, but when dealing with a complex system like physical asset management, we need to leverage some proven philosophical approaches to execute strategy effectively and deliver better results."

"How are we doing with time?" Darren stops to ask, staring at Vuhdam and Jerry alternately.

"Don't worry about time," Vuhdam replies instantly. "I think this is very important for all of us to understand, so that we can come up with the next best course of action."

"Perfect!" Darren says, rubbing his hands together with excitement.

Adopt a System Thinking

"I want you to always keep in mind that physical asset management is an organization-wide system with a complex network of interdependencies," Darren starts.

"How many of you have heard of Dr. W. Edwards Deming and his 'theory of profound knowledge'?" Darren then asks.

He scans the room and sees that most are nodding in agreement. To briefly explain, he says: "The theory of profound knowledge is a management philosophy grounded in systems theory. It is based on the principle that each organization is composed of a system of interrelated processes and people which make up the system's components. Dr. Deming stated that we should not outright blame people without having a closer look at our system and processes. He suggested that the manufacturing process is not a series of unrelated processes, but is an entire system, and, when viewed as an entire system, opportunities to improve efficiencies are more easily identified."

"This system thinking approach must be imperatively applied to the organization-wide physical asset management system for it to be effective. Deming offered fourteen key points for management to transform business effectiveness. Many, if not all of them, must be applied by organizations in their physical asset management approach," Darren says. And then pausing for a moment, he adds in a serious tone, "That is, if you really want a change and want to see positive results."

He looks at Vuhdam and turns to everyone in the room, explaining: "The dysfunctional silo approach that currently prevails in many organizations will not cut it anymore in physical asset management. A systemic approach is a must! I won't put anyone on the spot, but think about Mouroum Inc. to see if there is any symptom of functional silos and collaboration issues while you implement your physical asset management program."

"Darren, honestly from what I have been seeing over the years, I think we may have some of those issues," Vuhdam says.

Vuhdam looks at Jerry and continues: "I think this is something we have to address going forward. A systemic approach together with Deming's fourteen points can definitely help."

Jerry quickly jots down some notes to capture Vuhdam's comments.

Darren feels quite reinvigorated by the comments, and he takes advantage to move on to the next approach. He points to the second approach on the screen, saying, "Next we have to understand how value is realized from physical assets."

Understand the Physical Asset Management Value Chain

Darren begins: "In the present day, it is mind-boggling and hard to imagine an organization trying to achieve a goal without clearly understanding what that goal means and how it can be achieved."

Everybody in the room looks at Darren a little confused and surprised at the same time.

"This is somewhat the case with physical asset management," Darren continues right away. "In physical asset management we talk about 'value' and 'realizing value' quite often, but not enough is done to try to define what 'value' means."

He swiftly searches for and puts another slide on the screen (Figure 14.4).

ISO 55000 Definitions

ISO 55000 defines asset management as the "coordinated activities of an organization to realize value from assets."

ISO 55000 states that value (which can be tangible or intangible, financial or nonfinancial) will be determined by the organization and its stakeholders, in accordance with the organizational objectives.

OWPAM

FIGURE 14.4 ISO 55000 definitions.

"This is how 'value' is defined in ISO 55000," he resumes. "Now what about the process to realize value in physical asset management; has anyone ever considered the value chain involved and what it takes?"

Again, lots of people with surprised looks on there faces, all staring at him.

"Michael Porter, an academic from Harvard Business School, is well known for his theories on economics, business strategy, and social causes. He introduced the concept of the value chain. His idea of the value chain is based on the process view of organizations and understanding how value is created. The idea of looking at an organization as a system made up of subsystems each with inputs, transformation processes, and outputs is similar to how Dr. Deming describes it. Inputs, transformation processes, and outputs involve hundreds, even thousands, of activities which Michael Porter classifies as either primary or support activities that all businesses must undertake. How these activities in the transformation process are carried out determines the costs involved, which eventually affects profits. The Michael Porter value chain is a high-level model developed to describe the activities within and around an organization and evaluates the value each particular activity adds to its products or services. The primary activities are directly concerned with the creation of products or services, while the support activities help to improve effectiveness or efficiency."

"Many organizations develop a value chain to ensure that their transformation process is focused on delivering the required profit margin and that value flows into the transformation process."

By the reaction in the room, it looks like the concept of Michael Porter's value chain seems to resonate with everyone.

Darren follows up: "So why don't we develop a value chain for physical assets? Physical assets contribute significantly to the realization of profits, and there are many physical asset management activities involved that could affect those same profits."

Darren pauses, and turns to his laptop to pull up a diagram (Figure 14.5) on the screen. As people focus on the diagram, Darren continues: "In organization-wide physical asset management, the principle of the value chain is no different, except that in this case there are two value chains to consider—the physical asset lifecyle value chain and the functional area value chain."

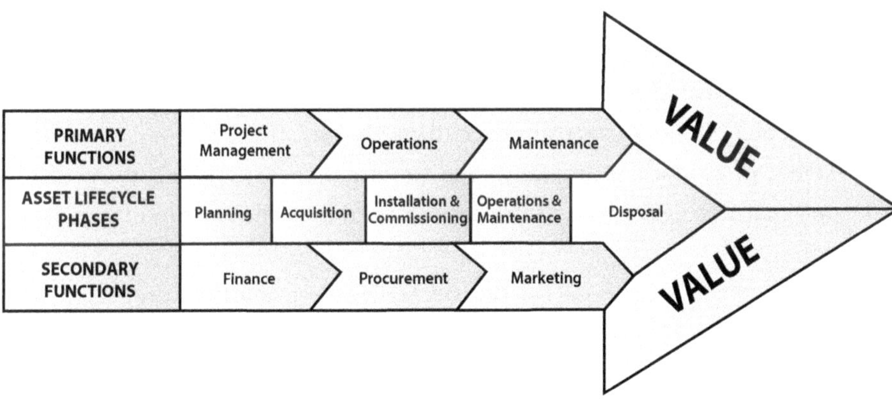

FIGURE 14.5. Physical asset management value chain.

"Can you explain what you have on the diagram please?" asks CIO Pravin Dimitrin, asks hesitantly.

"Yes, of course," Darren replies immediately.

"The physical asset life cycle phase value chain is what you see in the middle. This is where value is realized based on quality interrelated activities happening at different lifecycle phases of physical assets. These activities can take the form of policies, procedures, and processes that occur during the planning, acquisition, installation and commissioning, operations, maintenance, and disposal phases."

"The functional area value chain is where the different functions of the organization collaborate to maximize value. At the top of the chain, there are primary functions that are directly related to creating value from physical assets, such as operations, maintenance, and project management. Then there are the secondary functions at the bottom, which support the primary functions in achieving their goals, for example, procurement, finance, information technology, etc. Here the activities can be in the form of quality interactions, sharing of information, and coordination taking place between the different people involved in those functional areas."

"Thank you. This does make sense now," Pravin says in a more relaxed voice.

"As we have seen so far," Darren continues, "these two value chains are very important in the organizational physical asset management context and are interdependent upon each other to derive maximum value. Adding and preserving value in both value chains will achieve the vertical and horizontal alignment sought in an organization-wide physical asset management approach with the ultimate goal to realize maximum value."

"So, in practical terms, how do we realize value from the two value chains at Mouroum Inc.?" Vuhdam asks, almost cutting Darren off.

Darren, of course, understands the top guy's eagerness, but deep inside he knows that it is easier said than done and can imagine the amount of work involved.

"This is a valid question," Darren starts. "There is a lot of complexity and many intricacies of interdependencies that exist between the two value chains. Quality, involvement, and value flow are important to achieve effective alignment and coordination across the organization."

Darren starts to walk around the room. He is starting to feel at ease as if he is delivering a lecture to his students.

"First and foremost, we must recognize that many organizations may have a different functional area value chain depending on how the organization is structured and how the different functional areas are linked based on the type, size, and complexity of business activities. This will help identify the different components of the functional area value chain, as well as those of the physical asset lifecyle value chain."

"Does this mean that at Mouroum Inc. we will have to build our own functional area value chain?" Jerry asks, voicing a question for the first time.

"Yes, Jerry. It shouldn't be a tough process, as you just need to identify what functional areas should be primary functions and which ones are secondary functions," Darren answers.

"The next step is crucial," he continues. "This is when we have to identify the areas where there are opportunities to improve the value chains." And glancing at everyone around the table, Darren adds with a smile on his face, "This brings us to the theory of constraints approach."

Manage the Constraints Systematically

"I think I know about this approach," Robert says. "I attended a seminar many years ago where they introduced the methodology and I received a book titled *The Goal* by Dr. Eliyahu Goldratt—a very interesting book and approach, indeed."

Robert then turns to Vuhdam for a quick check. Darren is delighted to see Robert so engaged.

"That's excellent!" Darren says in an excited voice. "One of the reasons why organizations are struggling these days with physical asset management is that we look at it in functional isolation to make changes: add people here and there and develop processes and policies without realizing the impact on the other functional areas."

Darren takes a pause and then says in a stern voice: "This approach has to stop. Physical asset management is not about the performance of one or two functional areas such as finance or maintenance. It's about the performance of the value chain and all of its links."

Darren can feel the heavy silence in the room, and he continues: "To make changes and improve the performance of the physical asset management value chain, we need to focus on the weakest links of the chain, the constraints, and act on them" (Figure 14.6).

FIGURE 14.6 Weakest link of a value chain.

Darren explains: "The theory of constraints (TOC) is a management paradigm that views any manageable system as being limited in achieving more of its goals due to a small number of constraints. There is always at least one constraint, and TOC uses a focusing mechanism to identify the constraint and restructure the rest of the organization around it. TOC adopts the common idiom 'A chain is no stronger than its weakest link.' This means that processes, organizations, etc., are vulnerable, because the weakest part (or person) of a process can always damage or break them or at least adversely affect the outcome."

He stops, looks around, and further clarifies what he means: "When our goal is to realize maximum value from physical assets, we do not want anything in the process that will disrupt the realization of value or, in the worst case, destroy that value."

"The core concept of TOC is that every process has a single constraint and the total deliverable process can only be improved when the constraint itself is taken care of. Spending time and effort on non-constraints will not provide significant benefits; it is only when constraints are improved that more and more benefits are realized."

Darren skips through a few slides and then stops (Figure 14.7). Referring to the slide, he notes, "These are some key tools that are applied in the TOC philosophy. The five focusing steps are the cyclical process aimed at improving and eliminating constraints encountered in almost any process."

He then switches to the next slide (Figure 14.8).

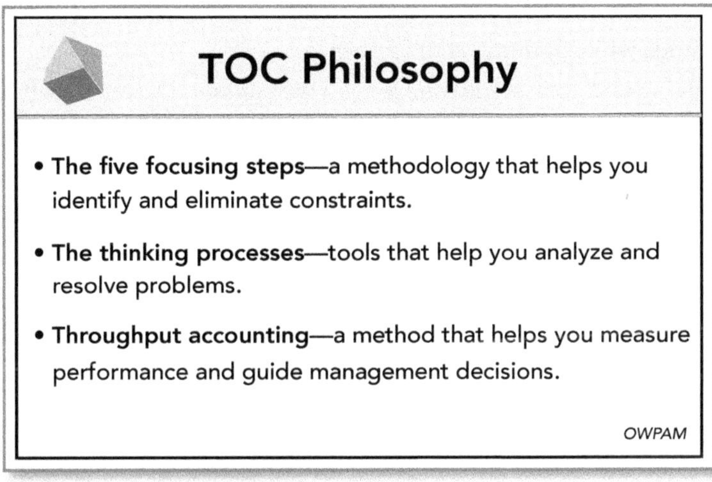

FIGURE 14.7 TOC philosophy.

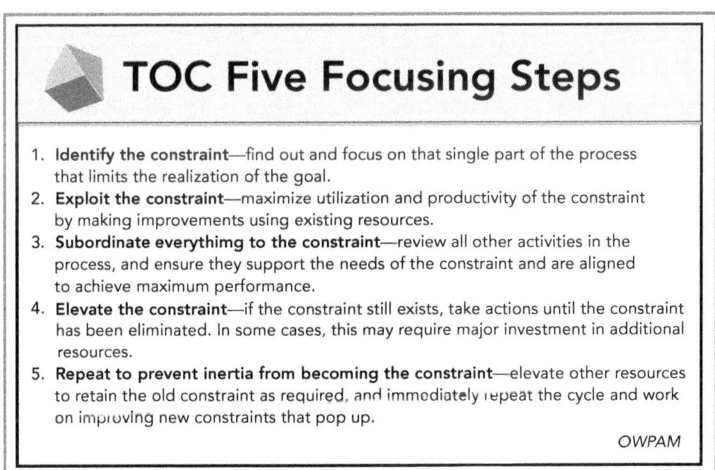

FIGURE 14.8 TOC five focusing steps.

"On this slide you can see what each step of the TOC five focusing steps entails," he adds while pointing at the screen.

After giving everyone some time to read the content, Darren flips back to the previous slide and continues: "The thinking processes help to answer three fundamental questions in the value chain.

He pauses and then adds, "We will talk about them in more detail later."

Then pointing to the third bullet in the slide, he says: "And lastly, throughput accounting. This is where the theory of constraints maintains the focus on realizing and increasing value through performance measures while making the changes."

Darren walks back to his seat and drinks a sip of water.

"All this sounds great and makes absolute sense," Mary starts. "However, I am interested in seeing some examples of how to apply the TOC to organization-wide physical asset management."

"Sure," Darren promptly replies. "First of all, let me explain the term 'constraint.' A constraint is anything that limits a system from achieving higher performance versus its goal. It is a step or process that is producing less than what is demanded of it."

He clears his throat and continues: "Constraints are those items that cause issues in our process or the physical asset management value chain to that effect. The constraints can be of different types, and some examples are listed in this table."

Darren scrolls to another slide (Figure 14.9). "Now let's take a look at a practical example of a constrained operations process. Let's say a lawnmower manufacturing company finds that its bottleneck operation is its assembly line. Assembly operations can only proceed at a certain pace, so the company can only manufacture forty lawnmowers per day. If the company were to produce more engines, the engines would not contribute to more lawnmowers being manufactured; there would only be an increase in the number of engines in inventory, which increases the cost of working capital. The CEO of the company finds that, since the number of lawnmowers produced per day is limited to forty, his next best course of action is to cut back production in all other areas if they are producing more parts than are needed for forty lawnmowers."

Constraints Examples

TYPE	EXAMPLE
Physical	Money, equipment, computer, human resources, facilities space
Policy	Regulations, contracts, procedures
Market	Demand, trends, politics
Paradigm	"We always refurbish those pumps every six months and replace all the components. (An instilled mindset or way of doing things.)

OWPAM

FIGURE 14.9 Constraints examples.

Darren stands up and starts walking back and forth at the front of the room. He cannot sit and speak at the same time. It's as if he gets his inspiration by moving around and, of course, gazing at the ceiling from time to time.

Distilling the point of the example, he says: "This shows that it is better to not optimize in many parts of the business, since the assembly line sets the pace, unless concrete action is taken to improve the assembly line. TOC provides an alternative view of individual or local productivity in a production line or organization. Just because individuals or departments can speed through tasks and create high outputs doesn't mean that they are benefiting the rest of the organization."

He takes a deep breath and continues: "Now let's see how we can apply this to organization-wide physical asset management and to its two value chains. Common constraints that can be found in a physical asset management system are lack of asset management awareness, organizational culture, shortage of skill sets, lack of resources, funding limitation, poor lifecycle management practices, inadequate governance and leadership, and many more. The more we drill down in the value chains of an organization, the more constraints can be uncovered within the different functional areas and at different levels of the organization."

Darren takes a few moments to gather his thoughts and then adds: "For example, in the physical asset lifecyle value chain, one of the constraints can be the installation and commissioning phase itself. When we drill down in the installation and commissioning value chain, we can uncover other underlying constraints that limit higher performance in the installation and commissioning of physical assets. Some of the constraints can be the design of the asset, the procurement policy, the prevailing project management practices, technical expertise, etc."

Darren takes a sip of water and quickly checks if everybody is following. Then he begins to talk again. "In the functional area value chain," he says, "we can follow the same approach. As an example, one of the constraints can be senior management. When we drill further down in this particular constraint, we can uncover other underlying constraints that act as roadblocks to achieving the strategic asset management goals. Some examples of this are the need of an asset management policy, lack of commitment, inconsistent decision making, poor leadership, and others."

Darren searches the faces of his listeners for any reaction to what he just said. He sees none and continues: "If we do not identify, acknowledge, and act on these constraints in a structured manner as outlined in the TOC, we'll never be able to realize maximum value for our value chains."

"This looks very interesting," interjects Vuhdam with some excitement in his voice. He looks at his team and says: "I think we should seriously consider these approaches to effectively focus on physical asset management. I see this as being no different from the lean principles that we applied on our operations. We define and understand what value means, we map the value chain together with performance measures, and we create that value flow by managing the bottlenecks in the process."

He sits up straight in his chair and says in an authoritative voice, "From now on, it will not be about who needs what physical asset investments, but primarily what the physical asset management value chain needs with a clear focus on realizing value."

Vuhdam then turns to Darren and asks, "What is the process to get all that to work at Mouroum Inc.? I presume it is not going to be a few days' job and we will most probably need many more people involved."

"This definitely is not going to be a few days' undertaking. Remember it is a transformation process which will significantly impact most, if not all, functional areas and their personnel. I suggest taking as much time as is needed; however, there must be sustained focus to derive maximum value from the organization-wide physical asset management philosophy," Darren responds.

"First and foremost," he adds, "everyone in an organization needs to understand how each one fits into the whole organization-wide physical asset management system. It's important to communicate its significance and work toward alignment, another one of the four fundamentals of physical asset management."

CHAPTER 15

OWPAM SYSTEMIC IMPLEMENTATION PROCESS

"Before we get to the process itself," Darren continues, as he looks at Vuhdam, "we need to prepare everyone to be receptive to adopting a system thinking approach."

"I like this approach," says Shika with a smile.

"I am somewhat confused about the 'system' you are referring to. Can you please elaborate on that a little bit?" Robert asks instantly.

"Good timing. We are actually going to look at the organization-wide physical asset management system right now."

Darren pulls up another slide (Figure 15.1). "This is the system we are looking for in an organization-wide physical asset management approach," he says as he points to the screen.

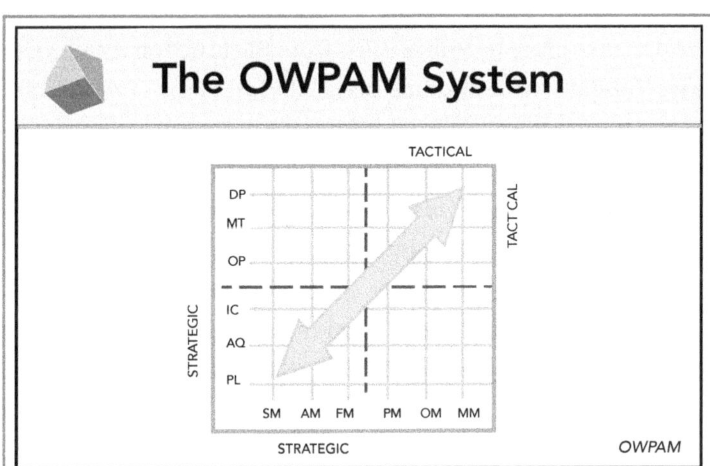

FIGURE 15.1 The OWPAM system.

Everyone is now busy staring at the screen trying to understand the diagram.

Darren waits a few minutes, and then he says: "This is the system that an organization operates to realize value from its physical assets and achieve its strategic physical asset management objectives. The goal is to manage the two value chains, the interrelation and interactions to achieve alignment."

Everybody is listening carefully to Darren while glancing back and forth at him and the screen.

Darren continues: "Let me take you through the diagram and provide further explanations."

Darren walks up to the screen and starts explaining: "The X-axis shows the functional areas and represents the functional area value chain. The Y-axis shows the different lifecycle phases and represents the physical asset lifecycle value chain."

"The grid represents the network of interdependencies existing in the complex system, where each intersection is a node. Each node is a combination of interactions and interrelations taking place in the functional area value chain and physical asset lifecycle value chain, respectively."

"Why do you have the functional areas in that order?" asks Vuhdam.

"Good question," replies Darren. "The way I lined up the functional areas is in the order from strategic to tactical going from left to right. You can add other functional areas as well depending on your organization. All of them form the interactions that need to take place to realize value from the functional area value chain."

"OK," says Vuhdam while nodding his head. "Is that the same case for the lifecycle phases on the Y-axis?"

"You bet," Darren answers. "The lifecycle phases are lined up from strategic at the bottom to tactical at the top. This represents the interrelations that must occur between the lifecycle phases in the physical asset lifecycle value chain."

Darren pauses and looks around to check if anyone has questions. Then he continues: "To derive maximum value from the two value chains, it is important to understand how they are interconnected and to consider the interactions and interrelations that take place. These take the form of a myriad of activities represented by a complex network of interdependencies."

Darren points at the nodes on the grid in the diagram and states in a serious tone, "The essence of an effective physical asset management system is to coordinate and align these activities to realize maximum value from the interaction and interrelation value chains."

After a few seconds of silence, Vuhdam says: "Darren, I have to admit that what you just explained makes really good sense and depicts clearly what an organization-wide physical asset management system is all about. Now I understand the complexity of the system and the importance of activities in both value chains."

"One more thing, if I may," interjects Darren. "The other important item I need to mention is about alignment. Too often the word 'alignment' is used in presentations and other documents; however, no thorough explanation is provided on what it actually is or how to implement it. Practitioners are left on their own to interpret what 'alignment' means for their organizations without any guidance on how to achieve it."

He continues, "As per the diagram, you can see that to achieve alignment in an organization-wide physical asset management system, there must be strategic to tactical alignment from both the interaction and interrelation perspectives, both at the functional areas level as well as the asset lifecycle phases level."

"This is not easy to achieve," says Robert.

"Absolutely," Darren instantly replies. "It is not an easy undertaking, as there are a lot of factors involved within such a complex system. In many organizations, alignment is achieved in a sparse manner, either in some functional silos alone or at some levels of the organization only, but not in a wider and consistent manner."

"What do you recommend then to achieve that consistency?" asks Vuhdam impatiently.

"I recommend taking a systemic thinking approach to provide the assurance of alignment across the whole organization. When we apply the systemic thinking processes methodology to complex systems with many interdependencies, as described in the theory of constraints, the systems can be optimized. In the case of physical asset management and in the two value chains, we need to identify the root causes of undesirable effects and then remove them without creating new ones."

Darren pulls up another slide (Figure 15.2). He explains, "The systemic thinking processes are used to answer the three questions essential to creating value and allowing it to flow."

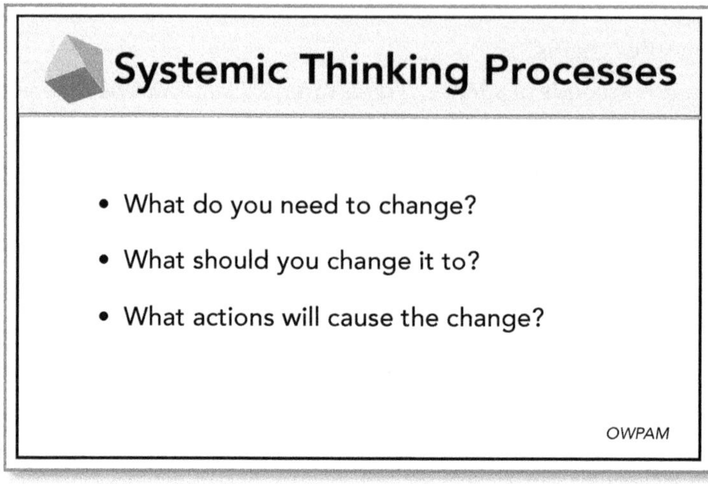

FIGURE 15.2 Systemic thinnking processes.

"At Mouroum Inc. you need to be aware that it is not just the activities in the functional areas that have to be considered in the value chains, but also the linkages between those activities and functions. The linkages are key elements in the chain that contribute to realize value. Each link can be broken down into processes/subprocesses to identify and map the value-producing activities. Similarly, if you look at the lifecycle phase activities, you have to consider the linkages between the activities and lifecycle phases to map the value-creating processes/subprocesses."

"Each link of the value chain is a source of constraint. The goal is to identify those constraints and improve them by taking appropriate actions. Different approaches such as continuous improvement programs, lean thinking, six sigma methodology, etc., are used with varying degrees of success. At the end, there will always be a constraint in the system, so organizations need to choose wisely which ones to act on and why. Eventually the system and its constraint should be stabilized to an acceptable level of performance and assurance."

Darren pauses to catch his breath and then goes on: "One caveat is that the systemic thinking process should not endorse frequently shifting constraints. This can create chaos and confusion with policies that are not properly enforced, processes that keep changing, and frequent staff turnover. This is the reason why a structured, well-planned, and documented systemic thinking approach is recommended to map out all the processes, subprocesses, policies, and any other changes required to manage the constraints effectively."

"What would that structured systemic thinking approach entail?" Vuhdam quickly asks, as Darren expected.

"Well, it is a multi-step process that may take multiple years to implement," Darren answers. "As you know by now, organization-wide physical asset management is a huge undertaking for any organization, as it entails a major transformation process."

Darren once again searches his laptop and pulls up another slide deck on the screen. He slowly scrolls down and stops to zoom in on a slide (Figure 15.3).

The OWPAM Systemic Thinking Steps

1. Define the goals of the OWPAM system and define value.
2. Establish the units of measurement of value.
3. Draw the OWPAM system and its value chains.
4. Assess the *as-is* of the OWPAM system.
5. Develop the *to-be* of the OWPAM system.
6. Identify the constraints in the system.
7. Prioritize the constraints in order of importance.
8. Develop an implementation plan.
9. Apply the TOC thinking process as per plan.
10. Implement the change while keeping the system stable.
11. Build the system and manage the constraint.
12. Create a suitable management structure.
13. Adopt a continuous improvement and learning culture.

OWPAM

FIGURE 15.3 The OWPAM systemic thinking steps.

"Steps 1 and 2 are part of your strategic asset management plan (SAMP), where you will identify and agree on the asset management objectives for Mouroum Inc.," Darren continues. "And Step 3 is where you will map your value chains based on your organizational context. Steps 4–8 are where you will assess the current situation, agree on what your ideal future state should be, and develop a plan to get there based on your priorities. Steps 9–13 are about acting on the constraints and managing the change over the long term, while continuously improving."

Darren looks up and asks, "Is everybody clear about the steps listed? Does anybody have any questions?"

Everybody seems to still be thinking while going down the list. Darren gives them another few seconds before he adds, "Steps 4–8 are very important. This is where you will identify gaps and overlaps between the as-is and to-be status and develop plans to tackle them."

Darren scrolls and stops on another slide listing some other steps (Figure 15.4).

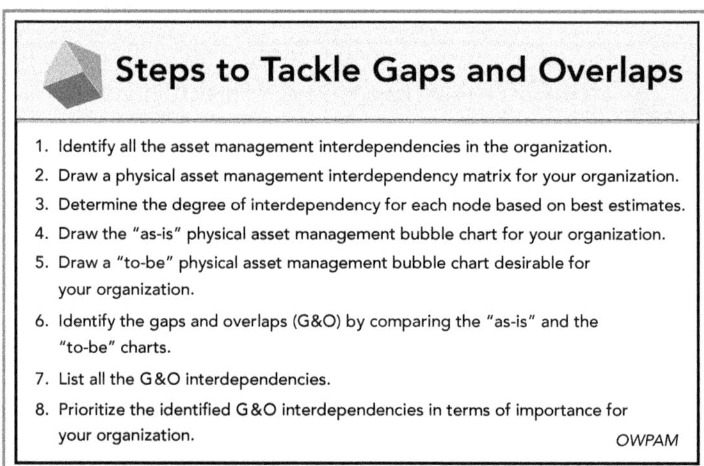

FIGURE 15.4 Steps to tackle gaps and overlaps.

Darren opens another document (Figures 15.5a and 15.5b) and then continues: "The two diagrams here show the as-is status of interdependencies existing in an organization and the desired to-be status."

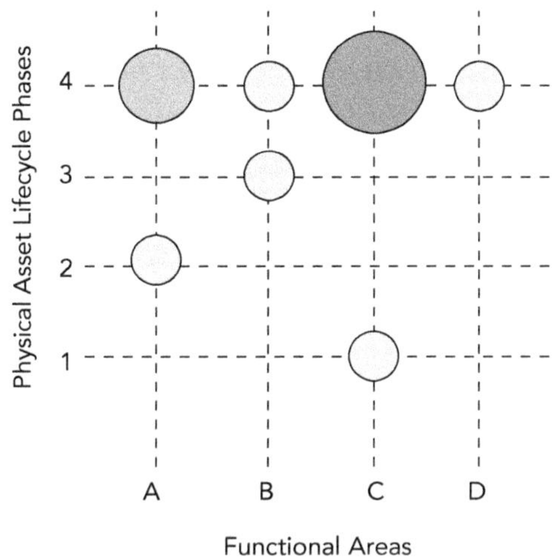

FIGURE 15.5a Example as-is status.

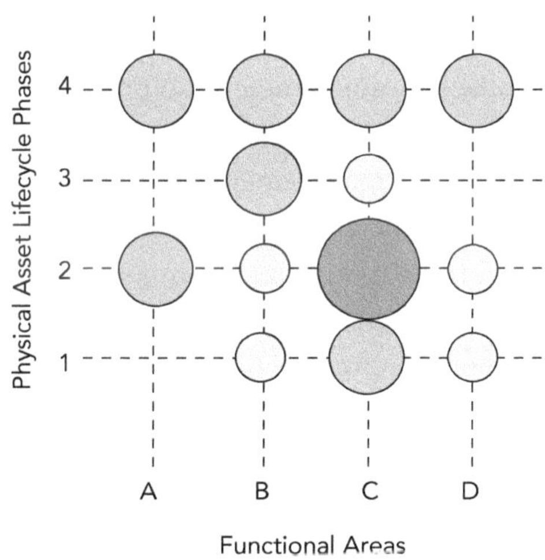

FIGURE 15.5b Example to-be status.

He picks up his pen and starts pointing at the diagrams, specifying, "In the 'as-is status' diagram, we can see potential gaps in lifecycle phases 1, 2, and 3. In lifecycle phase 4 there is potential for overlaps. The 'to-be status' diagram maps how to close those gaps and address the overlaps."

"To analyze the gaps let us look at interdependency C-1, for example. Suppose lifecycle phase 1 is the planning phase and functional area C is operations management. The 'as-is' status scenario (Figure 15.5a) shows that only operations management (C) is involved in the planning phase (1). The 'to-be status' scenario (Figure 15.5b) requires that other functional areas such as B and D be involved (interdependencies B-1 and D-1) while increasing the degree of interdependency in C-1. This can take the form of policies, processes, procedures, and activities that need to be identified, documented, and deployed for those interdependencies."

Darren looks around the table and notices some gloomy faces. Nevertheless, he continues his explanation. "To analyze the overlaps," he suggests, "let's look at lifecycle phase 4 and assume it to be the operations phase. In the 'as-is status' diagram, we can see the existing interdependencies A-4, B-4, C-4, and D-4, with interdependencies A-4 and C-4 being quite predominant. The 'to-be status' diagram recommends a more balanced network of interdependencies within the operations phase. To illustrate this, let's assume A is senior management, B is finance management, C is operations management, and D is asset management. The 'as-is status' diagram shows higher involvement of senior management and finance management in the operations phase, most probably to manage operations and meet the operational targets. The 'to-be status' diagram recommends increasing the involvement of finance management (B) and asset management (D) in the process while reducing the degree of interdependency of operations management. This can take the form of rolling out prioritization tools and procedures in a cross-functional team structure to develop more accurate operational and capital budgets taking everybody's inputs into consideration, while applying a consistent process throughout."

"This can be lots of work and overwhelming," Jerry replies in shock.

"Yes indeed," Darren agrees. "It will require lots of work, business process reengineering, process mapping, new procedures, change management, and all. The key is to identify where you want to be and acknowledge the gaps and overlaps. The rest can be planned and worked on over the years. Organizations need to be patient during this transformation process and implement it step-by-step using the appropriate strategy and resources. For example, once the organization knows its ideal future state, it can identify the top five interdependency priorities to start improving on."

Darren scrolls to another diagram (Figure 15.6). Focusing on the diagram, he says: "Here you can see the five top priorities to work on:

- Interdependencies B-1 and D-1
- Interdependency C-2
- Interdependencies C-4 and D-4"

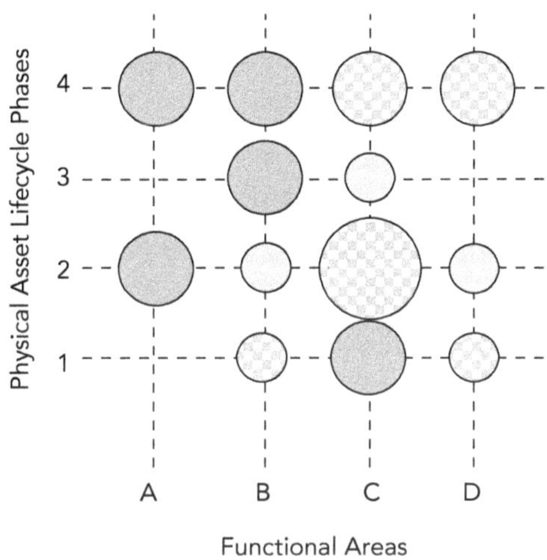

FIGURE 15.6 Top five interdependency priorities.

"At Mouroum Inc. you can put together a cross-functional working group to identify, develop, and implement policies, processes, and practices. This will help to improve collaboration, interactions, and sharing of information between those functional areas."

"Going from the as-is status to the to-be status is a major undertaking for any organization and is the most difficult phase in operationalizing a physical asset management system. However, we need to start doing things differently to change the current system. It is recommended that organizations take a careful and patient approach in the transformation process from the as-is state to the desired to-be state. Believe it or not, this is where the 'real value' of organization-wide physical asset management lies. This is where you are building an effective physical asset management system."

Darren stops and puts down the pen he has been holding.

"That's it from me for now," Darren says as he turns to Vuhdam. "Thank you for your time and for listening to me. Let me know if you have any questions."

"Thank you, Darren. It was great having you here to share your knowledge with us," declares Vuhdam.

He looks at Jerry and says: "Jerry, can you set up another meeting in a few weeks' time with all staff present here to discuss the next course of action?"

He then turns to his staff and continues: "I would like you to invite your VPs, directors, and managers as well. We need to meet and discuss the importance of implementing an organization-wide physical asset management system."

Everybody around the table nods affirmatively.

Jerry takes note of what Vuhdam just said and replies: "Yes, sir. I will check everybody's availability and schedule the meeting. Should I invite Darren as well?"

"No. Not for this meeting. I want to have an internal discussion first with staff," Vuhdam says calmly.

He turns to Darren and adds: "I would like to thank you so much for your time and insight on this complex topic of organization-wide physical asset management. We have been struggling for quite some time, and today has been an important eye-opener for us. It has really helped my staff and me to understand the magnitude of work ahead of us."

The he shakes Darren's hand and says with a smile: "You have done a brilliant job! I will ask Jerry to keep in touch with you because I plan to bring you back to Mouroum Inc. in the near future to guide us forward with this undertaking."

Vuhdam Jeragvika then leaves the room at a hasty pace similar to how he entered a few hours ago. Darren smiles and gives Jerry a quick look while he starts packing his stuff.

Darren feels pretty good as he attacks the last 3 miles of the Ironman 70.3. He has never felt better at this stage of a race, maybe because of his good preparation—strategically, with the plans he had in place and, of course, the full support and encouragement of his family.

He had a great swim at a very comfortable pace, just fast enough to keep up with the main group of swimmers. His bike split was not bad either, considering this is his biggest weakness. Luckily, he had no mechanical problems, one of the worst nightmares of any participant in a race. Darren finished the cycling portion in a reasonable time and tackled the running portion with enough energy left. For most of the run, he kept a conservative pace to ensure he felt good in his body and his mind, as he knows it will get harder toward the end. Darren has always had great admiration for the countless volunteers who are there along the whole course always ready to help and, with big smiles on their faces, always encouraging the participants.

Gradually Darren starts seeing more spectators on the sides of the road, all cheering louder and louder. He knows he is getting closer and closer to the finish line. He looks around knowing that Debbie, Rémi, and Lucas will be somewhere close by to cheer him on. Suddenly he hears his name from the crowd. He looks up and sees Jerry. What a surprise! Jerry is right there with Rita holding Mia alongside Debbie and the boys. Darren greets them with a huge smile and waves at them. As he comes closer, he gives them all a high five and then switches his focus to the finish line, which is some quarter mile away. He likes the sensation of running on the red carpet of an Ironman race with spectators applauding on both sides, the gigantic finish arch in front of him, and photographers all ready to get their best shots.

Nothing can describe the feeling of accomplishment when crossing the finish line, not only after such a tough race, but also after such hard, consistent, and long hours of training in the recent months.

As the sayings go: "Anything is possible" and "Pain is only temporary."

Darren knows this better than anyone else. Over the years he successfully applied these same principles of strategies and plans interchangeably to train for his Ironman race and to implement complex physical asset management systems in organizations to achieve desired goals. Now his biggest challenge remains applying the same approach and principles to get his doctoral degree in physical asset management.

INDEX

ABOUT THE AUTHOR

Dharmen Dhaliah, P. Eng., MBA, PMP, CAMA, MMP, CMRP, is a visionary and inspirational leader with a career-long record of strategic asset management, maintenance and reliability engineering, and project management success in complex organizations. He is a bilingual professional with proven talent for aligning business strategy and objectives with extensive industry expertise in leading the development and implementation of organization-wide asset management strategies in the public and private sectors. Dharmen is holder of a Master of Business Administration degree and is a Professional Mechanical Engineer, Certified Asset Management Assessor, Project Management Professional, Maintenance Management Professional, and Certified Maintenance & Reliability Professional. He is the published author of *Physical Asset Management: An Organizational Challenge*, and creator of the *Holistic Physical Asset Management (HPAM) Board Game*. Dharmen is an avid triathlete and an Ironman 70.3 finisher. He completed his first Ironman 70.3 at Lake Placid on September 9, 2018.

THE **HPAM** PROCESS

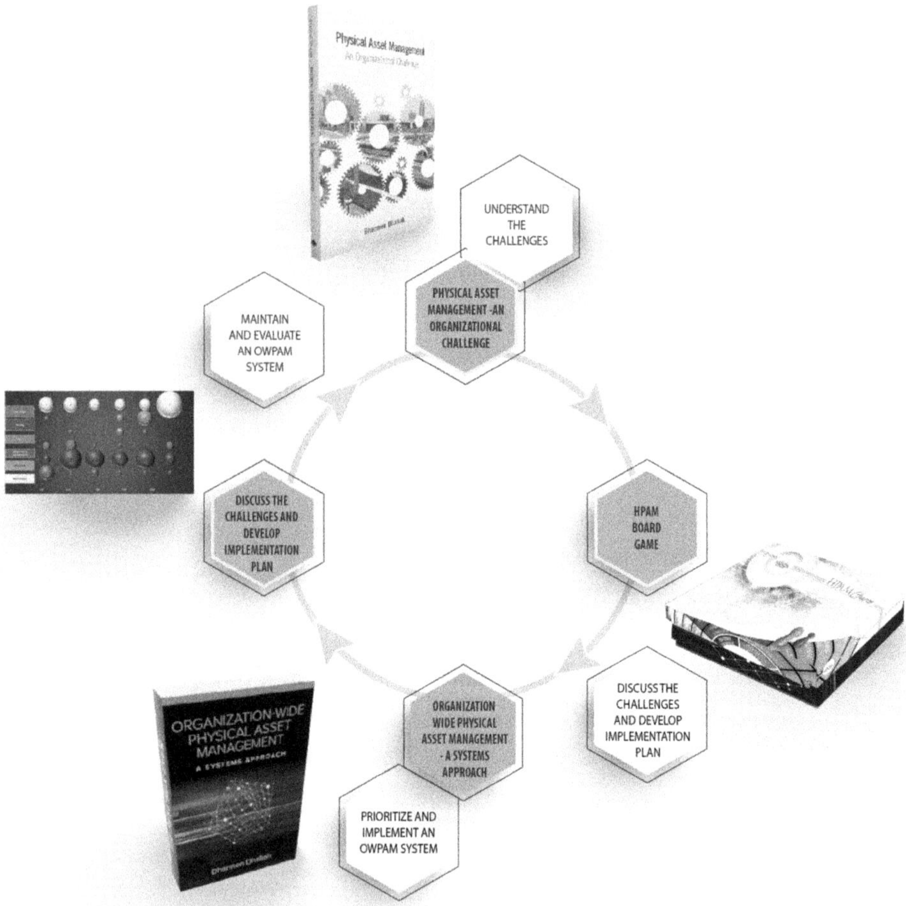

For more information visit www.hpamsolutions.com

www.ingramcontent.com/pod-product-compliance
Ingram Content Group UK Ltd.
Pitfield, Milton Keynes, MK11 3LW, UK
UKHW021825120926
8556IPUK00043B/1046

9 780831 136529